S0-ASW-713

OPTICAL PROPERTIES OF INHOMOGENEOUS MATERIALS

Applications to Geology, Astronomy, Chemistry, and Engineering

OPTICAL PROPERTIES OF INHOMOGENEOUS MATERIALS

Applications to Geology, Astronomy, Chemistry, and Engineering

Walter G. Egan
and
Theodore W. Hilgeman
Research Department
Grumman Aerospace Corporation
Bethpage, New York

ACADEMIC PRESS New York San Francisco London 1979
A Subsidiary of Harcourt Brace Jovanovich, Publishers

ACADEMIC PRESS, INC.
111 Fifth Avenue, New York, New York 10003

United Kingdom Edition published by
ACADEMIC PRESS, INC. (LONDON) LTD.
24/28 Oval Road, London NW1 7DX

Library of Congress Cataloging in Publication Data

Egan, Walter G
Optical properties of inhomogeneous materials.

Bibliography: p.
1. Inhomogeneous materials--Optical properties.
I. Hilgeman, Theodore W., joint author. II. Title.
QC173.4.I53E34 535 78-20043
ISBN 0-12-232650-4

PRINTED IN THE UNITED STATES OF AMERICA

79 80 81 82 9 8 7 6 5 4 3 2 1

CONTENTS

PREFACE

The intent of the authors in writing this book was, first of all, to clarify the newest studies of the optical properties of inhomogeneous materials and then to provide a guide to solve a number of related scientific and engineering problems based on these studies. These problems may come about in research endeavors ranging from remote sensing, aerial reconnaissance, study of the atmosphere, and energy collector design to the study of the surfaces provided by paints and other coatings. The fields of geology, optical mineralogy, astronomy, chemistry, soil mechanics, mechanical engineering, and of course optics are involved. At the same time we have tried to maintain the level of treatment in this book at that of undergraduate optics.

Several reasons account for this approach being taken now. There has been a great advance in the past twenty years in the techniques for experimentally and theoretically analyzing surface scattering by optical methods. A major impetus for this advance came from the Lunar Landing Program. There was a need to obtain as much information about the lunar soil as possible by remote characterization before the landing of a man on the moon. Remote characterization was required in order that the Lunar Module, as well as the lunar roving vehicles, could be designed efficiently. The lunar landing configurations were in great part influenced by terrestrial optical simulation of the lunar surface by spectrophotometric and polarimetric techniques. This work has been extended to the study of the surface and atmosphere of Mars, the earth's atmosphere, and the interstellar medium. Other factors catalyzing the advances in surface modeling were the prevalence of computer facilities having high speed computational capacity and the availability of instruments such as the scanning electron microscope with the ability to reveal surface microstructure.

This book may well be the first to present all the tools necessary for modeling the radiation scattered from diffuse surfaces. Included are the optical complex indices of refraction of a variety of naturally occurring minerals and rocks, appropriate mathematical surface models of varying complexity, and representative computer programs. The plan is a logical development starting with a brief introduction to the formalism for optical properties of inhomogeneous materials. This introduction (Chapter I) assumes that the reader has some familiarity with complex notation and vector algebra and, in particular, with their applications to electromagnetic radiation. This chapter also brings together in a unified formalism, a number of necessary reference formulas such as Snell's law in complex form. There follows a description of surface scattering models in order of increasing complexity and a discussion of atmospheric scattering by particulates (Chapter II). Then, using techniques based on Chapters I and II, the experimental approaches for the determination of the refractive and absorptive components of the optical complex indices of refraction are presented in Chapters III and IV, respectively, with sample indices in Chapter V.

We follow with actual diffuse surface modeling examples in Chapter VI, using complex indices from Chapter V. Retroreflectance, the unique property of a diffuse surface, is treated in Chapter VII. Then applications are discussed in succeeding chapters: remote sensing of planetary surfaces (Chapter VIII); study of the interstellar medium (Chapter IX); research on thermal energy collectors (Chapter X); examination of coatings and paints (Chapter XI); and remote mineral exploration (Chapter XII). The appendixes provide computer codes that may be used for investigations described in this book.

ACKNOWLEDGMENTS

We are indebted to many individuals for their contributions, but we wish to single out the Research Department of the Grumman Aerospace Corporation for providing an excellent intellectual environment in which many of the innovations described in this book were conceived and for providing the laboratory and computer facilities where these ideas could be developed and applied.

We thank the following Grumman Aerospace Corporation and Data Systems personnel: J. Augustine for his help in performing the optical measurements on the rocks and minerals discussed in Chapter V, some of which were previously unpublished; A. Kaercher, T. Mc Givney, and R. Reisch for rendering the graphs of Chapters II and V; A. Cobrin for his cooperation in arranging typing and art services; and especially R. Scheuing and M. D'Agostino for their encouragement and support.

We thank H. Holt, U.S. Geological Survey, Flagstaff, Arizona, for furnishing many of the samples measured for Chapter V.

We are also grateful for the patience shown by our wives during the many hours that we have had to devote to this book.

OPTICAL PROPERTIES
OF INHOMOGENEOUS MATERIALS

Applications to Geology, Astronomy,
Chemistry, and Engineering

CHAPTER I □ MATHEMATICAL FUNDAMENTALS

A. INTRODUCTION

The appearance of a surface to the eye, or to an optical measuring instrument, is a result of an involved series of processes at and within that surface. The appearance can vary with viewing direction and the nature of the illuminating radiation. The interaction of the illuminating radiation with the surface and body of a material is affected by the geometrical structure of the material and its basic optical properties. This text is concerned with this interaction, and the fundamental optical properties of materials which are described by the "optical complex index of refraction."

As a preparation for the optical analyses, we set forth in our notation the basic principles of the theory of electromagnetic radiation. For the purposes of this book, we treat the spectral range from the ultraviolet (0.18-μm wavelength) through the visible to the infrared (3-μm wavelength). The rationale for the restriction to this wavelength range is to exclude the necessity for discussion of radiative emission and energetic phenomena such as photoemission and x rays. Also, this is the transmission range for conveniently available optical materials (such as quartz) normally used for instrumentation in this wavelength range.

B. COMPLEX NOTATION

In order to introduce the mathematical notation and illustrate this interaction, we first shall discuss an example of simple reflection and absorption. Consider a homogeneous isotropic material with radiation incident upon it. The optical complex index of refraction $\hat{m}$ of a material is a complex number composed of two parts; a refractive (real) part and absorptive (imaginary) part,

$$\hat{m} = n - i\kappa_0 \tag{I-1}$$

The real part n is the ratio of the speed of electromagnetic radiation in a vacuum c to the speed v in the material,

$$n = c/v \tag{I-2}$$

The absorptive portion κ_0 is known as the extinction coefficient and arises from the fact that no substance is perfectly transparent. Even with the most transparent materials, a loss of intensity can be detected by using sufficient thicknesses.

If $\hat{I}_i$ is the intensity of a wave incident perpendicularly upon the surface of the uniform material medium, then an amount $\hat{I}_r$ will be reflected and will not enter; $\hat{I}_r$ is given by

$$\hat{I}_r = [(\hat{m} - 1)/(\hat{m} + 1)]^2 \hat{I}_i \tag{I-3}$$

where $\hat{m}$ is the optical complex index of refraction, and is generally a complex number. The physical significance of the complex nature of $\hat{I}_r$ is that the reflected wave undergoes both an amplitude change and a delay. The light that enters the medium is $\hat{I}_{i'}$, given by

$$\hat{I}_{i'} = \hat{I}_i - \hat{I}_r \tag{I-4}$$

Here too $\hat{I}_{i'}$ will be a complex number. The quantity $\hat{I}_i$ sets the time reference and may be arbitrarily taken to be real. One significance of the imaginary part of $\hat{I}_{i'}$ and $\hat{I}_r$ is in the calculation of interference effects; this will be described subsequently. We consider the absolute value of $\hat{I}_{i'}$ to represent the intensity of the radiation that enters the medium. Then, the intensity $\hat{I}_x$ at any depth x is given by Lambert's exponential law of absorption. Lambert's law of absorption states that layers of equal thickness, in a uniform material medium, absorb equal fractions of the intensity incident upon them. Thus,

$$\hat{I}_x = \hat{I}_{i'} e^{-\alpha x} \tag{I-5}$$

where the constant α is the absorption coefficient of the medium. It represents the fraction of incident intensity absorbed per unit thickness. The physical significance of the absorption coefficient α is that a portion of the radiation is absorbed to become heat (or other degraded energy forms). The absorption coefficient α is related to the extinction coefficient κ_0 of the optical complex index of refraction by the relation

$$\kappa_0 = \alpha\lambda/4\pi \tag{I-6}$$

where λ is the wavelength of the radiation. The origin of this expression will subsequently be shown in the discussion of Maxwell's equations.

C. SCATTERING

Lambert's exponential law of absorption is idealized because, in naturally occurring inhomogeneous materials, there is scattering of the incident radiation. In our usage, scattering occurs when the radiation is not permanently absorbed by the particles or molecules of a material, but absorbed and reradiated at essentially the same instant and wavelength. The result is that some of the incident radiation is deflected from the original direction into other directions including backward. Elementary textbooks frequently present the confusing and incorrect concept that an "absorption coefficient" can be associated with scattered radiation. In truth, the result for some specific examples is the same. However, such examples do not represent the complete scattering picture. There are other instances which we consider where absorption and scattering must be treated separately (see, e.g., Chapter II).

We alluded to the fact that most naturally occurring (and many man-made) rocks and minerals are inhomogeneous and scatter radiation. This means that their optical appearance will be the result of the optical complex index of refraction *and* the scattering properties. The scattering properties are caused by geometrical and structural properties of materials. These scattering properties can be caused by features ranging in size from very much smaller than the wavelength of the radiation to very much larger. These large-scale features can cause interparticulate shadowing effects.

The present theories of scattering (such as the Mie theory) require as input data the optical complex index of the material, free from scattering. The measurement of the optical complex index of refraction in naturally occurring inhomogeneous materials is fraught with pitfalls, and many incorrect data exist in the literature. The development of more-advanced scattering theories requires accurate and reliable optical complex indices of refraction. Further, certain analyses (dispersion relations) neglect the effect of scattering and are thus incorrect for scattering-type materials. The remainder of this chapter will be concerned with the determination of the optical complex index of refraction in the absence of scattering. Chapter II and those following will be concerned with scattering.

The mathematical formalism that forms the basis of the ensuing chapters will now be presented, together with a number of techniques that historically have been used for determining α and n. Their theoretical bases are presented with an evaluation of their accuracy. The more accurate techniques will be described in detail in Chapters III and IV.

D. BASIC PRINCIPLES OF ELECTROMAGNETIC RADIATION IN COMPLEX FORM

James Clerk Maxwell was the first to give a complete analytical description of the theory of electromagnetic radiation. The four equations constituting the theory may be expressed in a variety of ways, for instance, as two line integrals and two surface integrals, or as eight partial differential equations. An especially neat and compact form for a region without free charges makes use of vector notation. In free space, the magnetic permeability $\mu' = \mu_0 = 4\pi \times 10^{-7}$ H/m, and the dielectric constant $\epsilon' = \epsilon_0 = (1/36\pi) \times 10^{-9}$ F/m. However, in a conducting medium, where σ' = conductivity in (ohm meter)$^{-1}$, μ' and ϵ' are not necessarily equal to their free-space values.

$$\nabla \times \mathbf{E} = -\mu' \frac{\partial \mathbf{H}}{\partial t} \tag{I-7}$$

$$\nabla \times \mathbf{H} = \sigma' \mathbf{E} + \epsilon' \frac{\partial \mathbf{E}}{\partial t} \tag{I-8}$$

$$\nabla \cdot \mathbf{H} = 0 \tag{I-9}$$

$$\nabla \cdot \mathbf{E} = 0 \tag{I-10}$$

Using Eqs. (I-7) and (I-8) to eliminate **H**, and, since $\nabla \cdot \mathbf{E} = 0$, we obtain the wave equation

$$\nabla^2 \mathbf{E} = \mu' \epsilon' \frac{\partial^2 \mathbf{E}}{\partial t^2} + \mu' \sigma' \frac{\partial \mathbf{E}}{\partial t} \tag{I-11}$$

For monochromatic radiation of angular frequency ω $(= 2\pi\nu)$, the wave equation becomes

$$\nabla^2 \mathbf{E} + \hat{\mathbf{k}}^2 \mathbf{E} = 0 \tag{I-12}$$

where

$$\hat{\mathbf{k}}^2 = \omega^2 \mu' (\epsilon' + i\sigma'/\omega) \tag{I-13}$$

Note that the wave number $\hat{\mathbf{k}}$ is a complex number, containing the magnetic permeability μ' of the medium and a complex dielectric constant $\hat{\epsilon}'$, where

$$\hat{\epsilon}' = \epsilon' + i\sigma'/\omega \tag{I-14}$$

The quantities μ', ϵ', and σ' may be tensors, as in crystal field theory, but for inhomogeneous materials they appear in the scattering terms.

The quantity μ' may also be complex for a lossy magnetic material. In analogy to Eq. (I-1), let us define the optical complex index of refraction $\hat{m}$ in terms of the complex wave number $\hat{\mathbf{k}}$:

$$\hat{\mathbf{k}} = \omega \hat{m} \tag{I-15}$$

$$\hat{m} = n(1 - i\kappa) \tag{I-16}$$

Comparing with Eq. (I-1), we see that $\kappa_0 = n\kappa$. Substituting (I-15) and (I-16) into (I-13) we find that

$$n^2 = +\tfrac{1}{2}[(\mu'^2\epsilon'^2 + \mu'^2\sigma'^2/\omega^2)^{1/2} + \mu'\epsilon'] \tag{I-17}$$

and

$$n^2\kappa^2 = +\tfrac{1}{2}[(\mu'^2\epsilon'^2 + \mu'^2\sigma'^2/\omega^2)^{1/2} - \mu'\epsilon'] \tag{I-18}$$

The relationship between σ' and κ is clarified by noting that for the case of the pure dielectric $\sigma'/\omega = 0$, and then $n = (\mu'\epsilon')^{1/2}$ and $\kappa = 0$.

A simple solution of the wave equation is an exponential representing an attenuated sinusoidal plane wave progressing in the x direction,

$$\mathbf{E} = \mathbf{E}_0 \exp[-i\omega t + i\hat{\mathbf{k}}(x/c)] \tag{I-19}$$

$$= \mathbf{E}_0 \exp[-i\omega t + i\omega\hat{m}(x/c)] \tag{I-20}$$

$$= \mathbf{E}_0 \exp[-i\omega t + i\omega n(x/c) - \omega n\kappa(x/c)] \tag{I-21}$$

The term $\exp(-\omega n\kappa x/c)$ represents the attenuation of the plane-wave amplitude with distance. The energy-density decrease is proportional to the square of this term, and, in a distance $c/2\omega n\kappa$, the energy density falls to $1/e$. This defines the absorption coefficient α,

$$\alpha = 2\omega n\kappa/c \tag{I-22}$$

$$= 4\pi\nu n\kappa/c \tag{I-23}$$

$$= 4\pi n\kappa/\lambda \tag{I-24}$$

It is to be emphasized that the foregoing derivation does not include the effect of scattering.

We now come to the derivation of Fresnel's equations from Maxwell's equations. These equations give the reflection and refraction of a plane wave incident on a plane surface.

For the plane transverse wave progressing in the x direction given by Eq. (I-19), vector amplitude $\mathbf{E}$ (transverse to the x direction) can be equivalently considered to be the vector sum of two orthogonal components in a plane perpendicular to the x direction,

$$\mathbf{E} = \mathbf{E}_y + \mathbf{E}_z \tag{I-25}$$

In free space, the selected directions y and z are arbitrary, and, by a suitable selection of axes, one or the other may be made zero. In such a case, the wave would be plane polarized in the nonzero direction. In the general case, for such a wave incident on an arbitrarily oriented surface (which locates a plane of incidence containing the propagation vector and the normal to the surface), there will exist two components $\mathbf{E}_y = \mathbf{E}_{i\parallel}$ (incident electric field parallel to the plane of incidence) and $\mathbf{E}_z = \mathbf{E}_{i\perp}$ (incident electric field perpendicular to the plane of incidence); these components will be assumed to be in time phase at incidence and real, as with Eq. (I-4). This assumption is made to simplify our analysis. The more-complicated cases of elliptically or circularly polarized radiation can be derived by including the effect of the phase difference between $\mathbf{E}_{i\parallel}$ and $\mathbf{E}_{i\perp}$ in the linear polarization expressions presented.

The boundary conditions require that at the surface the normal components of $\epsilon'\mathbf{E}$ and $\mu'\mathbf{H}$ and the tangential components of $\mathbf{E}$ and $\mathbf{H}$ be continuous. Relative to the surface normal, a wave incident at angle $\hat{\theta}_i$ is refracted at angle $\hat{\theta}_{i'}$. (Note that $\hat{\theta}_i$ and $\hat{\theta}_{i'}$ are considered to be complex numbers; in general, the significance of a complex angle is that the imaginary portion changes an unattenuated propagating wave into an attenuated one.) Then,

$$\omega_z t - \hat{\mathbf{k}}_z \sin\hat{\theta}_i = \omega_{z'} t - \hat{\mathbf{k}}_{z'} \sin\hat{\theta}_{i'} \tag{I-26}$$

where z' refers to the second medium. This is true for all t; thus,

$$\omega_z = \omega_{z'} \tag{I-27}$$

and

$$\hat{\mathbf{k}}_z \sin\hat{\theta}_i = \hat{\mathbf{k}}_{z'} \sin\hat{\theta}_{i'} \tag{I-28}$$

or [using (I-15)]

$$\omega_z \hat{m}_i \sin\hat{\theta}_i = \omega_{z'} \hat{m}_{i'} \sin\hat{\theta}_{i'} \tag{I-29}$$

$$\hat{m}_i \sin\hat{\theta}_i = \hat{m}_{i'} \sin\hat{\theta}_{i'} \tag{I-30}$$

where $\hat{m}_i$ and $\hat{m}_{i'}$ are the complex indices of refraction of the incident and refractive media. Equation (I-30) is Snell's law of refraction at a surface in complex form. Note that the use of some purely refractive techniques to determine the index of refraction (such as the method of minimum deviation using a prism) is incorrect in the presence of absorption.

For the case of a conducting medium having radiation incident from a dielectric medium, a useful form of Snell's law utilizing real angles is

$$n \sin\theta_i = n'(\theta_i) \sin\theta_{i'} \tag{I-31}$$

Here, the effective index $n'(\theta_i)$ of the conducting medium is a function of the incident angle θ_i and is given by

$$n'(\theta_i) = (n^2 \sin^2\theta_i + q^2)^{1/2} \tag{I-32}$$

where

$$q^2 = \tfrac{1}{2}\{n'^2 - \kappa_0'^2 - n^2 \sin^2\theta_i + [4n'^2\kappa_0'^2 + (n'^2 - \kappa_0'^2 - n^2\sin^2\theta_i)^2]^{1/2}\} \tag{I-33}$$

Similarly, for the reflected beam angle $\hat{\theta}_r$, one finds

$$\hat{\theta}_r = \hat{\theta}_i \tag{I-34}$$

The reflected and refracted waves relative to the plane of incidence may be expressed in vector form. When the conditions of continuity across the boundary are applied and μ is assumed the same in both media, we obtain the amplitude of the refracted beams,

$$\mathbf{E}_{i'\|} = \mathbf{E}_{i\|}\left|\frac{2\sin\hat{\theta}_{i'}\cos\hat{\theta}_i}{\sin(\hat{\theta}_i + \hat{\theta}_{i'})\cos(\hat{\theta}_i - \hat{\theta}_{i'})}\right| \tag{I-35}$$

$$\mathbf{E}_{i'\perp} = \mathbf{E}_{i\perp}\left|\frac{2\sin\hat{\theta}_{i'}\cos\hat{\theta}_i}{\sin(\hat{\theta}_i + \hat{\theta}_{i'})}\right| \tag{I-36}$$

and the amplitude of the reflected beam

$$\mathbf{E}_{r\|} = +\mathbf{E}_{i\|}\left|\frac{\tan(\hat{\theta}_i - \hat{\theta}_{i'})}{\tan(\hat{\theta}_i + \hat{\theta}_{i'})}\right| \tag{I-37}$$

$$\mathbf{E}_{r\perp} = -\mathbf{E}_{i\perp}\left|\frac{\sin(\hat{\theta}_i - \hat{\theta}_{i'})}{\sin(\hat{\theta}_i + \hat{\theta}_{i'})}\right| \tag{I-38}$$

When the reflected beam component $\mathbf{E}_{r\|}$ equals zero (and this can happen only when $\hat{\theta}_i$ and $\hat{\theta}_{i'}$ are real, i.e., when there is no absorption), the reflected wave present is polarized completely perpendicular to the plane of incidence. The emergent angle, termed the Brewster angle θ_B, is obtained by setting $\mathbf{E}_{r\|}$ in (I-37) equal to zero.

This implies that $\theta_i + \theta_{i'} = \pi/2$; now, using Eq. (I-30), it is readily shown that

$$\tan\theta_B = \hat{m}_{i'}/\hat{m}_i = n_{i'}/n_i \tag{I-39}$$

For dielectric or very low absorption materials, the Brewster angle determination can be used to determine the refractive portion of the complex index of refraction. However, for absorptive materials, $\mathbf{E}_{i'\|}$ [Eq. (I-37)] does not go to zero at any angle, but it does have a minimum. For values of $\kappa_0 < 0.01$, the determination of the real index, using Eq. (I-39) with θ_B as the angle of the minimum, is accurate to three significant figures. We now turn to the reflectivities from a surface implied by Eqs. (I-37) and (I-38). The reflectivities for the rays incident perpendicular to the plane of incidence ($R_\perp$) and those parallel to the plane of incidence ($R_\|$) are given by the squares of the amplitudes. These expressions are obtained by expanding the tangent terms in sines and cosines and using Snell's law [Eq. (I-30)].

$$\hat{\mathbf{R}}_\| = \left(\frac{\mathbf{E}_{r\|}}{\mathbf{E}_{i\|}}\right)^2 = \left\{\frac{(\hat{m}_{i'}/\hat{m}_i)^2\cos\hat{\theta}_i - [(\hat{m}_{i'}/\hat{m}_i)^2 - \sin^2\hat{\theta}_i]^{1/2}}{(\hat{m}_{i'}/\hat{m}_i)^2\cos\hat{\theta}_i + [(\hat{m}_{i'}/\hat{m}_i)^2 - \sin^2\hat{\theta}_i]^{1/2}}\right\}^2 \tag{I-40}$$

$$\hat{\mathbf{R}}_\perp = \left(\frac{\mathbf{E}_{r\perp}}{\mathbf{E}_{i\perp}}\right)^2 = \left\{\frac{\cos\hat{\theta}_i - [(\hat{m}_{i'}/\hat{m}_i)^2 - \sin^2\hat{\theta}_i]^{1/2}}{\cos\hat{\theta}_i + [(\hat{m}_{i'}/\hat{m}_i)^2 - \sin^2\hat{\theta}_i]^{1/2}}\right\}^2 \tag{I-41}$$

$\hat{\mathbf{R}}_{\perp}$ and $\hat{\mathbf{R}}_{\parallel}$ are, in general, complex. The advantage of this procedure is that we have now expressed the reflectivities in terms of the usually known quantities, the indices of refraction and the incident angle.

For ellipsometry, we shall need the azimuthal angle $\hat{\psi}_r$ of the reflected light defined by

$$\tan^2 \hat{\psi}_r = \hat{\mathbf{R}}_{\parallel}/\hat{\mathbf{R}}_{\perp} = R^2 \exp(-2i\Delta \tan^2 \hat{\psi}_i) \quad \text{(I-42)}$$

where $\hat{\psi}_i$ is defined by

$$\tan \hat{\psi}_i = \mathbf{E}_{i\parallel}/\mathbf{E}_{i\perp} \quad \text{(I-43)}$$

The quantity Δ represents the phase shift between the parallel and perpendicular reflected components, and R is the ratio of their amplitudes. (Note that we are continuing the assumption of linearly polarized incident radiation.)

When $\mathbf{E}_{r\perp}$ and $\mathbf{E}_{r\parallel}$ are in time phase, the reflected radiation is plane polarized, $\Delta = 0$, and $\hat{\psi}_r$ is real. This is always true for pure dielectrics but occurs only at grazing incidence in the general case. If the time phase between $\mathbf{E}_{r\perp}$ and $\mathbf{E}_{r\parallel}$ differs by $\frac{1}{2}\pi(=\Delta)$, the reflected radiation is elliptically polarized, except if $|\mathbf{E}_{r\perp}| = |\mathbf{E}_{r\parallel}|$ when it is circularly polarized. When $\Delta = \frac{1}{2}\pi$, the corresponding incident angle is called the principal angle of incidence. At this angle, linearly polarized light is reflected in such a manner that the axes of the vibration ellipse are parallel and perpendicular to the plane of incidence. For normal incidence, $\Delta = \pi$ and the outgoing radiation is again plane polarized. Only for the principal angle of incidence is it possible to introduce a phase compensator (retardation plate) into the reflected beam in such a way as to transform the elliptically polarized beam into a linearly polarized output. It is this property that forms the basis for the technique of ellipsometry which will be discussed in Chapter IV.

Note that all the foregoing derivations deal with nonscattering materials. The inclusion of scattering involves differing degrees of modification of the foregoing basic theory.

E. EFFECT OF SCATTERING ON REFLECTIVITY

We now consider the applicability of the Brewster angle technique and ellipsometry in the presence of scattering. Scattering is caused by optical inhomogeneities of materials, such as crystal lattice dislocations, polycrystalline faces, particulate inclusions, heterogeneous composites, and anisotropies, to name a few. The effect of scattering is to redirect the incident radiation without loss. Absorption losses occur when the radiation passes through an additional depth of material as a result of scattering.

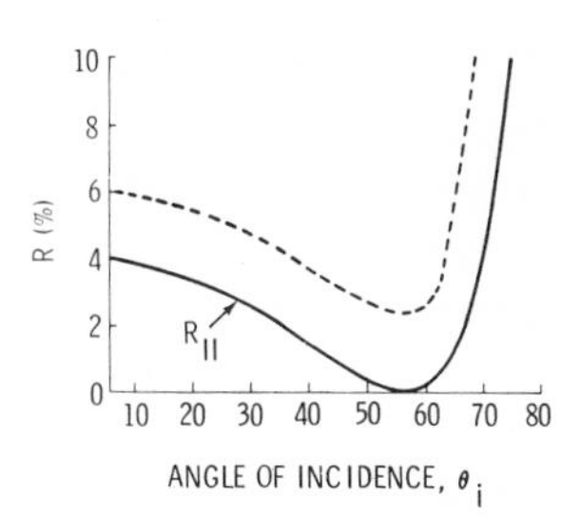

FIG. I-1 Reflectivity as a function of incidence angle for $n = 1.5$ without scattering, i.e., Fresnel, (——) and with scattering (– – –); note that scattering may be much larger than shown.

Consider an idealized dielectric of index of refraction 1.5 ($\hat{m}_{i'}/\hat{m}_i$). The parallel reflected component goes to zero at the Brewster angle (Fig. I-1). However, when scattering is present, the minimum will still occur, but it will not go to zero (Fig. I-1). If the scattered contribution does not vary with angle of incidence, the minimum will not be shifted; however, in practice, the scattering usually causes a slight shift to larger angles because the effect of scattering is least at grazing incidence. This effect can be minimized by requiring a high polish on the surface of the samples. In practice, real indices of refraction accurate to three significant figures can be obtained using the Brewster angle technique, even if scattering losses approach 90% of the incident beam. The techniques of ellipsometry are also effective for scattering materials because the polarization selectivity of the apparatus discriminates against the scattering.

We now turn to a discussion of techniques for the determination of the absorption coefficient. The absorption coefficient α is easily determined in nonscattering materials by measuring the decrease in intensity of a transmitted beam in the material, allowing for reflection at the incident and exit faces.

Another approach to the determination of n and κ_0 is to measure reflectivity at two angles; or to measure $\mathbf{R}_\perp$ and $\mathbf{R}_\parallel$ at the same angle of incidence. A generalized approach of this type using reflectivity at normal incidence alone is the Kramers–Kronig, or dispersion-relation analysis. For instance, the reflectivity, amplitude r, and the phase shift ϕ as a function of wavelength for the parallel component (which for normal incidence equals the perpendicular component) can be expressed as [from Eq. (I.37)],

$$\mathbf{E}_{r\parallel}/\mathbf{E}_{i\parallel} = re^{i\phi} \tag{I-44}$$

where $r = |\hat{\mathbf{R}}|^{1/2}$. The Kramers–Kronig relation links $\ln r$ to ϕ,

$$\phi(\nu_0) = \frac{2\nu_0}{\pi}\int_0^\infty \frac{\ln r}{\nu^2 - \nu_0{}^2}\,d\nu \tag{I-45}$$

The integral is evaluated at frequency ν_0 and requires a knowledge of r at all wavelengths. Referring again to Eq. (I-40), we see that a knowledge of ϕ and r allows the computation of n and κ_0.

This approach does not allow for scattering in the medium, and serious errors can occur if one attempts to use a Kramers–Kronig analysis for a scattering medium. Briefly, the reflectivity is changed and the phase information of Eq. (I-45) is lost. Analyses appropriate to scattering media are described in Chapter II.

F. SUMMARY

Important optical notation and relationships for use later in the book (e.g., $\hat{m}$, α, and $\hat{\mathbf{R}}$) were introduced. The expressions were developed further using classical approaches, and a number of practical relationships were derived in complex form. The application of these relationships to measurement of the optical complex index of refraction was discussed, and some of the effects of scattering were considered.

NOTES ON SUPPLEMENTARY READING MATERIAL*

Abraham and Becker (1950): Presents the concepts of vectors, vector fields, the electric and the magnetic field, and energy and forces in Maxwell's theory in a readily understandable form.

Born and Wolf (1959): Presents a reasonably complete picture of our present knowledge of optics; in particular, of those optical phenomena that may be treated in terms of Maxwell's equations.

Frank (1966): A compact logical exposition of the fundamental electric and magnetic field laws with applications to optics and the electrical and magnetic properties of matter.

Jenkins and White (1957): Classical physical optics with emphasis on the physical explanation of optical phenomena using graphical means; the complementary aspects of wave and particle properties of radiation are treated in a unified way.

Kramers (1927): see Kronig (1938).

Kronig (1938): Descriptions of a technique for the determination of optical constants of materials from reflectivity measurements.

Richtmyer *et al.* (1955): Classical and modern discussion of the origin, evolution, and present status of some important concepts in physics; considers electromagnetic waves and optical spectra.

Robinson (1952): see Robinson and Price (1953).

* Complete references are given at the back of this book.

Robinson and Price (1953): Describes a generalization of a technique for the determination of the optical constants of materials from reflectivity measurements in terms of dispersion relations.

Seitz (1940): Presents the physics that deals with the electronic structure of solid bodies, both theoretical and experimental aspects.

Slater and Frank (1947): A development of electromagnetism from first principles, with appendixes containing sufficient mathematical background that familiarity with calculus and differential equations permits following the derivations.

Stratton (1941): Treats variable electromagnetic fields and theory of wave propagation from fundamental concepts to later application; requires a general knowledge of electricity and magnetism.

CHAPTER II □ THEORETICAL SCATTERING MODELS THAT INCLUDE ABSORPTION

A. INTRODUCTION

The properties of the radiation scattered from or passing through a medium are dependent upon the optical complex index and upon geometrical factors. Physically, the medium may be a solid, a liquid, or a gas. The prediction of the scattered radiation requires the use of an appropriate model. Certain models are better than others; in this chapter we will describe seven models, in order of increasing detail, that include absorption and scattering to describe the extinction of a beam of light that enters a medium. Extinction is defined as the amount of radiation removed from a beam by absorption and the scattering processes. Increased accuracy in prediction of the scattered radiation generally involves a more complicated model and, concomitantly, more computer calculation. Selected computer programs are given in the Appendixes for several useful approaches.

The first four models to be described are more commonly applicable to solids (the two-flux model, the two-flux model with surface reflection, the six-flux model, and the modified dispersion model). The three models following these are generally used for atmospheric scattering (the Mie scattering, radiative transfer, and multiple-scattering models). The radiative transfer model is quite general, however, and may be applied to solids as well. We do not separately consider scattering in liquids, although the same principles are applicable.

The energy transfer process will now be discussed in general. Energy transfer through a medium is accomplished by electromagnetic waves. These waves are described by Maxwell's equations [(I-7) to (I-10)]. This may be shown as follows: If we take the dot product of $\mathbf{H}$ with Eq. (I-7) and the negative dot product of $\mathbf{E}$ with Eq. (I-8), we obtain

$$\mathbf{H}\cdot(\nabla\times\mathbf{E})-\mathbf{E}\cdot(\nabla\times\mathbf{H})$$
$$=-\mu'\mathbf{H}\cdot\frac{\partial\mathbf{H}}{\partial t}-\epsilon'\mathbf{E}\cdot\frac{\partial\mathbf{E}}{\partial t}-\sigma'\mathbf{E}\cdot\mathbf{E}=\nabla\cdot(\mathbf{E}\times\mathbf{H}) \tag{II-1}$$

noting that

$$\nabla\cdot(\mathbf{E}\times\mathbf{H})=\mathbf{H}\cdot(\nabla\times\mathbf{E})-\mathbf{E}\cdot(\nabla\times\mathbf{H}) \tag{II-2}$$

and, if ϵ' and μ' are constants, then

$$\epsilon'\mathbf{E}\cdot\frac{\partial\mathbf{E}}{\partial t}=\frac{\epsilon'}{2}\frac{\partial(\mathbf{E}\cdot\mathbf{E})}{\partial t}=\frac{\epsilon'}{2}\frac{\partial(\mathbf{E}^2)}{\partial t} \tag{II-3}$$

and

$$\mu'\mathbf{H}\cdot\frac{\partial\mathbf{H}}{\partial t}=\frac{\mu'}{2}\frac{\partial(\mathbf{H}\cdot\mathbf{H})}{\partial t}=\frac{\mu'}{2}\frac{\partial(\mathbf{H}^2)}{\partial t} \tag{II-4}$$

Then, by integrating over a volume V in a region where the electromagnetic waves exist, we obtain an expression describing the distribution of energy within and leaving the volume. Thus,

$$\int_v\left[\mu'\frac{\partial}{\partial t}\left(\frac{\mathbf{H}^2}{2}\right)+\epsilon'\frac{\partial}{\partial t}\left(\frac{\mathbf{E}^2}{2}\right)+\sigma'\mathbf{E}^2\right]dV=\int_v\nabla\cdot(\mathbf{E}\times\mathbf{H})\,dV \tag{II-5}$$

By the divergence theorem (Gauss's theorem)

$$\int_v\nabla\cdot\mathbf{F}\,dV=\int_s\mathbf{F}\cdot d\mathbf{s} \tag{II-6}$$

the total internal energy change from a closed region may be obtained by taking the surface integral over the surface enclosing the region. Thus,

$$\int_v\left[\frac{\partial}{\partial t}\left(\frac{\mu'\mathbf{H}^2}{2}\right)+\frac{\partial}{\partial t}\left(\frac{\epsilon'\mathbf{E}^2}{2}\right)+\sigma'\mathbf{E}^2\right]dV=-\int_s(\mathbf{E}\times\mathbf{H})\cdot d\mathbf{s} \tag{II-7}$$

The terms $\mu'\mathbf{H}^2/2$ and $\epsilon'\mathbf{E}^2/2$ represent the energy stored in the electromagnetic and electrostatic fields within the volume, and $\sigma'\mathbf{E}^2$ represents the Joule heat dissipated within the volume by free charge conduction. The term on the right-hand side represents the rate of energy flow into the volume, or, changing sign, the rate of energy flow out:

$$W=\int_s\mathbf{P}\cdot d\mathbf{s} \tag{II-8}$$

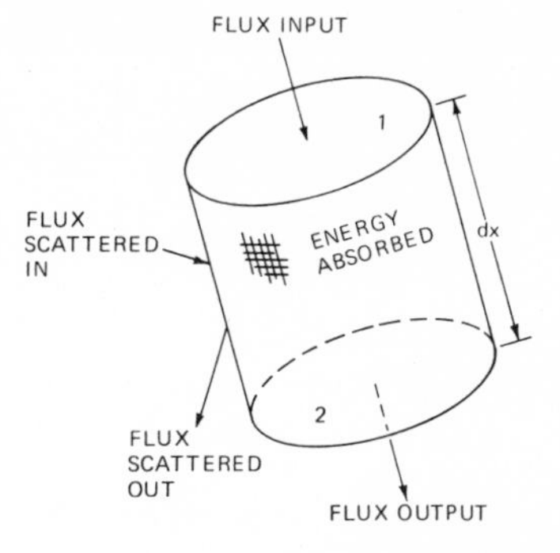

FIG. II-1 Elementary volume element containing radiant energy, with flux entering and leaving.

where

$$\mathbf{P} = \mathbf{E} \times \mathbf{H} \tag{II-9}$$

and **P** is termed the Poynting vector, which represents a flux.

Essentially, the scattering models that we will use are based on the consideration of a small volume in space of arbitrary shape through which radiant energy is passing. The radiant energy, in passing through this volume, may suffer a net change. The net change is represented by the difference in flux crossing two elementary surfaces (Fig. II-1). From conservation of energy, the difference in the flux crossing surfaces 1 and 2 is accounted for by flux scattered into and out of the side walls of the volume element and also by energy absorbed or created within the volume. As mentioned earlier, we are restricting this book to systems having no energy sources within the volume.

On this basis, the fundamental equation of radiative transport in this book with reference to an infinitesimal volume is given by

$$\begin{aligned}(\text{net flux change}) &= (\text{flux input}) - (\text{flux continuing out}) \\ &= (\text{flux scattered out}) + (\text{energy absorbed}) \\ &\quad -(\text{flux scattered in})\end{aligned} \tag{II-10}$$

Note that a similar relationship applies to intensity (flux per unit solid angle) as applies above to flux.

B. SCATTERING MODELS FOR SOLIDS

1. Two-Flux Model [Kubelka–Munk (KM) Theory]

The Kubelka–Munk scattering theory is based on the solution of simultaneous differential equations of the first order in one dimension. That is, the model assumes plane symmetry for a unit cross section.

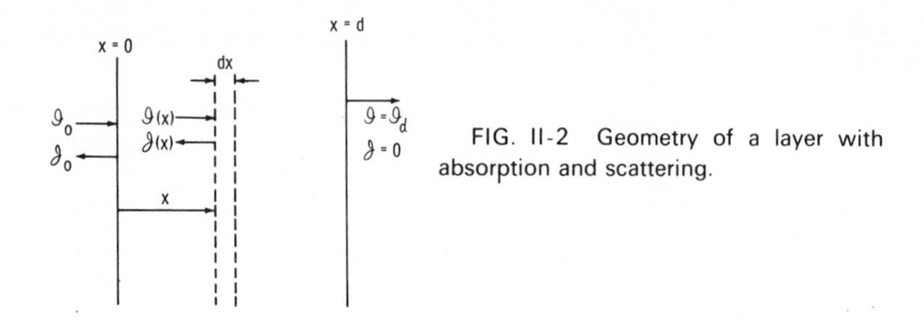

FIG. II-2 Geometry of a layer with absorption and scattering.

Consider Fig. II-2: the radiant flux in the positive x direction is $\mathscr{I}$, while that returned (as a result of scattering) is $\mathscr{J}$. At a distance x from the surface, across an infinitesimal thickness dx, Eq. (II-10) becomes

$$d\mathscr{I} = -(\alpha + \sigma)a_0 \mathscr{I}\, dx + \sigma b_0 \mathscr{J}\, dx \tag{II-11}$$

$$d\mathscr{J} = +(\alpha + \sigma)b_0 \mathscr{J}\, dx - \sigma a_0 \mathscr{I}\, dx \tag{II-12}$$

where α is the fraction of radiation absorbed per unit path length, σ is that fraction of radiation scattered per unit path length, and a_0 and b_0 are constants that relate dx to the average path lengths for the radiation $\mathscr{I}$ and $\mathscr{J}$. (Here, σ should not be confused with σ', the conductivity.) For ideal diffusers, the scattered radiation has the same intensity in all directions and $a_0 = b_0 = 2$.

The differential equations are then

$$d\mathscr{I} = -(k + s)\mathscr{I}\, dx + s\mathscr{J}\, dx \tag{II-13}$$

$$d\mathscr{J} = +(k + s)\mathscr{J}\, dx - s\mathscr{I}\, dx \tag{II-14}$$

where $k = 2\alpha$ and $s = 2\sigma$. The solution of these equations is an exponential, and, by applying the boundary conditions for the layer (Fig. II-2), we obtain the expressions for total diffuse reflection R_{KM} and transmission T_{KM} for diffuse incident radiation,

$$R_{\mathrm{KM}} = \frac{(1 - \beta^2)(e^{Kd} - e^{-Kd})}{(1 + \beta)^2 e^{Kd} - (1 - \beta)^2 e^{-Kd}} \tag{II-15}$$

$$T_{\mathrm{KM}} = \frac{4\beta}{(1 + \beta)^2 e^{Kd} - (1 - \beta)^2 e^{-Kd}} \tag{II-16}$$

where

$$K = [k(k + 2s)]^{1/2} \tag{II-17}$$

and

$$\beta = [k/(k + 2s)]^{1/2} \tag{II-18}$$

For an infinitely thick sample, the transmission is zero and the reflectivity [Eq. (II-15)] becomes

$$R_\infty = \frac{1-\beta}{1+\beta} = \frac{1-[\alpha/(\alpha+2\sigma)]^{1/2}}{1+[\alpha/(\alpha+2\sigma)]^{1/2}} \tag{II-19}$$

These equations are not readily inverted, but may be solved by iteration on k and s when R_{KM} and T_{KM} are the measured quantities. Diffuse incident radiation is assumed, but usually, in spectroscopy, the incident radiation is collimated. In some cases the equations may be valid: for homogeneous diffusers, collimated light incident at 60° to the surface normal will render them valid. These equations are useful for analyses of powders and their size distributions.

2. Two-Flux Model with Surface Reflection and Collimated Incidence [Modified Kubelka–Munk (MKM) Theory]

a. Derivation of the Model

The modified Kubelka–Munk (MKM) model was developed by Reichman (1973) as a more accurate solution of the scattering problem that takes into account the usual spectroscopic condition of collimated incident radiation. Simple anisotropy was included in the model. The explicit expressions for transmittance and reflectance were derived using the Schuster–Schwartzchild approximation to the equation of transport. We will outline the approach.

Consider a uniform dispersion of scatterers in a parallel-plane-bounded medium. The cosine of the angle between a reference direction and the direction of increasing thickness is denoted by μ. Distance in the medium is given by the dimensionless parameter τ where $\tau = (\alpha + \sigma)x$, and α, σ, and x have the same significance as in the Kubelka–Munk model. The dimensionless parameter τ is termed the "optical depth" and is more convenient for radiative transfer calculations than the physical depth. In order to describe collimated incident radiation in a manner that permits a reasonably simple solution for the radiative transfer problem, an artifice is employed: The incident radiation striking the surface at $\cos^{-1} \mu_i$ is represented by a conical shell making an angle $\cos^{-1} \mu_i$ to the surface normal; the intensity is now independent of the azimuthal angle φ by symmetry, and this simplification reduces the internal scattering problem to variables μ and μ_s, the cosines of the directions of the incident and scattered radiation to the surface normal. The intensity of the collimated radiation in the thin shell can be described by the function $\mathscr{I}_0 \delta(\mu - \mu_i)$, where δ is the Dirac delta function, and $\mathscr{I}_0$ is the incident intensity.

The equation of transfer for the scattered diffuse radiation field is given by Reichman (1973) as

$$\mu \frac{d\mathcal{I}(\mu)}{d\tau} = -\mathcal{I}(\mu) + \frac{\omega}{4\pi}\mathcal{I}_0 e^{-\tau/\mu_i} \int_{-1}^{1}\int_{0}^{2\pi} p(\mu, \varphi; \mu_s, \varphi_s)\delta(\mu - \mu_i)\, d\mu_s\, d\varphi_s$$

$$+ \frac{\omega}{4\pi}\int_{-1}^{1}\int_{0}^{2\pi} \mathcal{I}(\mu_s)p(\mu, \varphi; \mu_s, \varphi_s)\, d\mu_s\, d\varphi_s \qquad \text{(II-20)}$$

where $p(\mu, \varphi; \mu_s, \varphi_s)$ denotes the scattering phase function for the angle between the directions specified by (μ, φ) and (μ_s, φ_s). The term on the left-hand side denotes the change in intensity as the diffuse radiation passes through the medium. The first term on the right-hand side applies to the radiation absorbed into or scattered out of the region. The next term is a pseudo-source function which has been added to account for the existence of the initial beam of collimated incident radiation that is being absorbed and scattered as it progresses through the medium; the term was added by Reichman (1973) to improve the analytical representation. The last term refers to diffuse radiation scattered into the region. This expression is a direct application of Eq. (II-10). The single-scattering albedo $\tilde{\omega}$ is given by:

$$\tilde{\omega} = \sigma/(\alpha + \sigma) \qquad \text{(II-21)}$$

The intensity has been made independent of azimuth angle, and at the angle $\varphi = 0$ the integrations yield

$$\mu \frac{d\mathcal{I}(\mu)}{d\tau} = -\mathcal{I}(\mu) + \tilde{\omega}\mathcal{I}_0 e^{-\tau/\mu_i}F(\mu, \mu_i) + \tilde{\omega}\int_{-1}^{1} F(\mu, \mu_s)\mathcal{I}(\mu_s)\, d\mu_s \qquad \text{(II-22)}$$

where

$$F(\mu, \mu_s) = \frac{1}{4\pi}\int_{0}^{2\pi} p(\mu, 0, \mu_s, \varphi_s)\, d\varphi_s \qquad \text{(II-23)}$$

Now we use the Schuster–Schwartzchild approximation [i.e., the intensities are isotropic in the forward hemisphere ($\mu > 0$) and backward hemisphere ($\mu < 0$)]. If the forward intensity is $\mathcal{I}$ and the backward intensity is $\mathcal{J}$, the following equations are obtained by integrating Eq. (II-22):

$$\frac{1}{2}\frac{d\mathcal{I}}{d\tau} = -(1 - \tilde{\omega}F_d)\mathcal{I} + \tilde{\omega}(1 - F_d)\mathcal{J} + \tilde{\omega}\mathcal{I}_0 F_c(\mu_i)e^{-\tau/\mu_i} \qquad \text{(II-24)}$$

$$-\frac{1}{2}\frac{d\mathcal{J}}{d\tau} = -(1 - \tilde{\omega}F_d)\mathcal{J} + \tilde{\omega}(1 - F_d)\mathcal{I} + \tilde{\omega}\mathcal{I}_0 e^{-\tau/\mu_i}[1 - F_c(\mu_i)] \qquad \text{(II-25)}$$

where F_d is the fraction of radiation that is scattered in the forward hemisphere by a scatterer with incident diffuse illumination and $F_c(\mu_i)$ is the corresponding fraction for the collimated radiation from direction μ_i. The forward-scattering fractions are

$$F_d = \int_0^1 \int_0^1 F(\mu, \mu_s)\, d\mu\, d\mu_s \tag{II-26}$$

$$F_c(\mu_i) = \int_0^1 F(\mu, \mu_i)\, d\mu \tag{II-27}$$

The phase function can be expanded in a series of Legendre polynomials for the azimuth-independent case to give

$$F(\mu, \mu_s) = \frac{1}{2} \sum_{s=0}^{\infty} a_s P_s(\mu) P_s(\mu_s) \tag{II-28}$$

Substituting (II-28) into (II-26) and (II-27), we find

$$F_d = \frac{1}{2}\left[1 + \sum_{s=1}^{\infty} a_s g_s^2\right] \tag{II-29}$$

$$F_c(\mu_i) = \frac{1}{2}\left[1 + \sum_{s=1}^{\infty} a_s g_s P_s(\mu_i)\right] \tag{II-30}$$

where

$$g_0 = 0, \qquad g_1 = \tfrac{1}{2}, \qquad \text{and} \qquad g_s = [(2 - s)/(1 + s)]g_{s-2} \tag{II-31}$$

Solving (II-24) and (II-25), we find

$$\mathscr{I} = Ae^{-\gamma\tau_1} + Be^{\alpha\tau_1} - G\mathscr{I}_0 e^{-\tau_1/\mu_i} \tag{II-32}$$

$$\mathscr{J} = R_\infty Ae^{-\gamma\tau_1} + (B/R_\infty)e^{\alpha\tau_1} - GH\mathscr{I}_0 e^{-\tau_1/\mu_i} \tag{II-33}$$

where $\tau_1 = (\alpha + \sigma)l$, l is the geometric thickness of medium,

$$\gamma = 2(1 + 2\tilde{\omega}^2 F_d - \tilde{\omega}^2 - 2\tilde{\omega}F_d)^{1/2} \tag{II-34}$$

$$G = \frac{4\tilde{\omega}(F_c + \tilde{\omega} - \tilde{\omega}F_c - \tilde{\omega}F_d + F_c/2\mu_i)}{1/\mu_i{}^2 - \gamma^2} \tag{II-35}$$

and

$$H = \frac{1 - F_c + \tilde{\omega}F_c - \tilde{\omega}F_d - (1 - F_c)/2\mu_i}{F_c + \tilde{\omega} - \tilde{\omega}F_c - \tilde{\omega}F_d + F_c/2\mu_i} \tag{II-36}$$

The diffuse reflectance of a semi-infinite slab for diffuse incidence is

$$R_\infty = (2 - 2\tilde{\omega}F_{\mathrm{d}} - \gamma)/2(1 - F_{\mathrm{d}})\tilde{\omega} \tag{II-37}$$

The quantity R_∞ [Eq. (II-37)] is equivalent to the Kubelka–Munk expression [Eq. (II-19)] when we set

$$\sigma_{\mathrm{KM}} = 2\sigma(1 - F_{\mathrm{d}}) \tag{II-38}$$

The parameters A and B arise from the boundary conditions

$$\mathscr{I}(0) = 0 \tag{II-39}$$

$$\mathscr{J}(\tau_1) = 0 \tag{II-40}$$

Substituting (II-39) and (II-40) into (II-32) and (II-33), we determine A and B. If we use the following equations for the reflectance and transmittance

$$R = \mathscr{J}(0)/2\mu_{\mathrm{i}}\mathscr{I}_0 \tag{II-41}$$

$$T = (\mathscr{I}(\tau_1)/2\mu_{\mathrm{i}}\mathscr{I}_0) + e^{-\tau_1/\mu_{\mathrm{i}}} \tag{II-42}$$

we obtain

$$R = \frac{G[R_\infty - H(1 - R_\infty^2)\exp[-(\tau_1/\mu_{\mathrm{i}} + \gamma\tau_1)] - H - R_\infty(1 - R_\infty H)\exp(-2\gamma\tau_1)]}{2\mu_{\mathrm{i}}[1 - R_\infty^2\exp(-2\gamma\tau_1)]} \tag{II-43}$$

$$T = \frac{G(1 - R_\infty^2)\exp(-\gamma\tau_1) - (1 - R_\infty H)\exp(-\tau_1/\mu_{\mathrm{i}}) + R_\infty(R_\infty - H)\exp[-(\tau_1/\mu_{\mathrm{i}} + 2\gamma\tau_1)]}{2\mu_{\mathrm{i}}[1 - R_\infty^2\exp(-2\gamma\tau_1)]} + \exp(-\tau_1/\mu_{\mathrm{i}}) \tag{II-44}$$

These equations give the diffuse reflectance and transmittance for collimated incident radiation on a slab in terms of absorption and scattering coefficients and the phase function.

b. Effect of Reflecting Boundaries

Internal and external surface reflections must be considered if the refractive index of the scattering medium differs from unity. Assume that specular reflection occurs. For collimated incident light, the surface reflectance is given by Fresnel's equations. The radiation incident on the internal surface is partially diffuse and partially specular. Orchard (1969) has tabulated the internal reflectance for uniformly diffuse incidence as a function of refractive index. For a highly anisotropic phase function, the internal diffuse reflectance should be considered to be an experimentally determined parameter.

The apparent reflectance and transmittance for collimated incident radiation may be derived using Orchard's method,

$$\mathscr{R} = \frac{(1 - r_i)(1 - r_e)[r_i T_d T' + r_i r_e T_d R e^{-\tau_1/\mu_i} + (R + r_e T' e^{-\tau_1/\mu_i})(1 - r_i R_d)]}{(1 - r_e^2 e^{-2\tau_1/\mu_i})[(1 - r_i R_d)^2 - r_i^2 T_d^2]} + \frac{r_e + r_e(1 - 2r_e)e^{-2\tau_1/\mu_i}}{1 - r_e^2 e^{-2\tau_1/\mu_i}} \quad \text{(II-45)}$$

$$\mathscr{T} = \frac{(1 - r_i)(1 - r_e)[(T' + r_e R e^{-\tau_1/\mu_i})(1 - r_i R_d) + r_i T_d(R + r_e T' e^{-\tau_1/\mu_i})]}{(1 - r_e^2 e^{-2\tau_1/\mu_i})[(1 - r_i R_d)^2 - r_i^2 T_d^2]} + \frac{(1 - r_e)^2 e^{-\tau_1/\mu_i}}{1 - r_e^2 e^{-2\tau_1/\mu_i}} \quad \text{(II-46)}$$

where r_e is the Fresnel equation specular reflectance [Eqs. (I-40) and (I-41)] and

$$r_i = \text{internal diffuse surface reflectance} = 1 - (1 - r_s)/n^2 \quad \text{(II-47)}$$

$$r_s = \text{external diffuse surface reflectance} \cong 0.4399 + 0.7099n - 0.3319n^2 + 0.0636n^3 \quad \text{(II-48)}$$

[curve fit to tabular data of Orchard (1969)]. The symbols R and T' refer to the reflectance and transmittance for collimated incident radiation given by Eqs. (II-43) and (II-44) except that the exponential transmittance term is omitted. R_d and T_d are the reflectance and transmittance with diffuse incident radiation,

$$R_d = R_\infty(1 - e^{-2\gamma\tau_1})/(1 - R_\infty^2 e^{-2\gamma\tau_1}) \quad \text{(II-49)}$$

$$T_d = (1 - R_\infty^2)e^{-\gamma\tau_1}/(1 - R_\infty^2 e^{-2\gamma\tau_1}) \quad \text{(II-50)}$$

where γ and R_∞ are given by Eqs. (II-34) and (II-37). Note that $R_d = R_{KM}$ [Eq. (II-15)] if $\gamma\tau_1 = Kd$; and, similarly, $T_d = T_{KM}$ [Eq. (II-16)].

c. Isotropic Scattering

For isotropic scattering, including the effects of surface reflection with collimated normally incident radiation, we have

$$T_i = \frac{[(T_c + r_0 R_c e^{-\tau})(1 - r_i R_D) + r_i T_D(R_c + r_0 T_c e^{-\tau})](1 - r_0)(1 - r_i)}{(1 - r_0^2 e^{-2\tau})[(1 - r_i R_D)^2 - r_i^2 T_D^2]} + \frac{(1 - r_0)^2 e^{-\tau}}{1 - r_0^2 e^{-2\tau}} \quad \text{(II-51)}$$

$$R_i = \frac{[(R_c + r_0 T_c e^{-\tau})(1 - r_i R_D) + r_i T_c T_D + r_i r_0 T_D R_c e^{-\tau}](1 - r_i)(1 - r_0)}{(1 - r_0^2 e^{-2\tau})[(1 - r_i R_D)^2 - r_i^2 T_D^2]} + \frac{(1 - r_0)^2 r_0 e^{-2\tau}}{1 - r_0^2 e^{-2\tau}} + r_0 \quad \text{(II-52)}$$

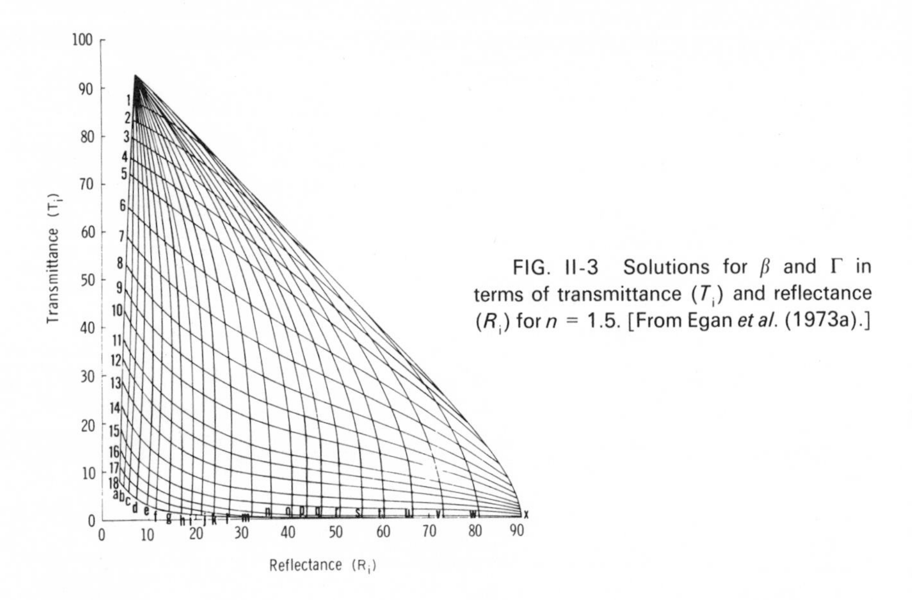

FIG. II-3 Solutions for β and Γ in terms of transmittance (T_i) and reflectance (R_i) for $n = 1.5$. [From Egan *et al.* (1973a).]

where T_i is the total measured transmittance for collimated normal incident illumination, isotropic scattering, R_i the total measured reflectance for collimated normal-incident illumination, isotropic scattering, τ the $(\alpha + \sigma)d$, and r_0 the normal-incidence surface reflection, i.e.,

$$r_0 = [(n - 1)/(n + 1)]^2 \tag{II-53}$$

$$T_c = \frac{(1 - \beta^2)}{(8\beta^2 - 2)} \frac{(R_\infty - 3R_\infty{}^2)e^{-(2\Gamma+\Gamma)} + (3 - R_\infty)e^{-\tau} + 3(R_\infty{}^2 - 1)e^{-\Gamma}}{1 - R_\infty{}^2 e^{-2\Gamma}} \tag{II-54}$$

$$R_c = \frac{(1 - \beta^2)}{(8\beta^2 - 2)} \frac{(R_\infty{}^2 - 1)e^{-(\tau+\Gamma)} + e^{-2\Gamma}(3R_\infty - R_\infty{}^2) + 1 - 3R_\infty}{1 - R_\infty{}^2 e^{-2\Gamma}} \tag{II-55}$$

$$R_D = \frac{R_\infty(1 - e^{-2\Gamma})}{1 - R_\infty{}^2 e^{-2\Gamma}} \qquad (= R_{KM} \quad \text{when} \quad \Gamma = \gamma\tau_1) \tag{II-56}$$

and

$$T_D = \frac{(1 - R_\infty{}^2)e^{-\Gamma}}{1 - R_\infty{}^2 e^{-2\Gamma}} \qquad (= T_{KM} \quad \text{when} \quad \Gamma = \gamma\tau_1) \tag{II-57}$$

TABLE II-1 Index to Curves on Fig. II-3[a]

Index	Γ	Index	β	Index	Γ	Index	β
1	0.1	a	1.0	13	2.3	m	0.28
2	0.2	b	0.96	14	2.7	n	0.24
3	0.3	c	0.86	15	3.2	o	0.20
4	0.4	d	0.76	16	3.7	p	0.18
5	0.5	e	0.68	17	4.2	q	0.16
6	0.7	f	0.60	18	5.0	r	0.14
7	0.9	g	0.54			s	0.12
8	1.1	h	0.48			t	0.10
9	1.3	i	0.44			u	0.08
10	1.5	j	0.40			v	0.06
11	1.8	k	0.36			w	0.04
12	2.0	l	0.32			x	0.02

[a] From Egan *et al.* (1973a).

In Eqs. (II-54) and (II-55),

$$\Gamma = 2[\alpha(\alpha + 2\sigma)]^{1/2}d \tag{II-58}$$

and

$$\beta = [\alpha/(\alpha + 2\sigma)]^{1/2} \tag{II-59}$$

with α the absorption coefficient, σ the scattering coefficient, and d the thickness of the specimen. In Eqs. (II-56) and (II-57), R_D and T_D are the diffuse reflectance and transmission with diffuse illumination.

For example, for a medium with $n = 1.5$, graphical solutions for Γ and β are shown in Fig. II-3 [from which α and σ are determined by Eqs. (II-58) and (II-59)] in terms of the experimentally observable quantities T_i and R_i. Table II-1 lists the index to the parameters β and Γ. The increments in β and Γ were chosen in order to space the curves relatively uniformly. Locate the intersection point for the measured transmittance and reflectance on the curves; this point determines the values Γ and β; interpolate where necessary.

3. Six-Flux Model

This model was developed by Emslie and Aronson (1973) to take into account the effect of particle shape and the effect of coherent interaction between closely spaced fine particles in a particulate scattering medium. Features such as edges and surface asperities cause absorption and as such are included in the theory. However, the scattering effect of edges and asperities was not included in the original version; we shall include a description of an appropriate modification of the original theory in Section II, 4.

There are two parts to the six-flux theory: one version treats coarse particles and the other version treats fine particles. The coarse-particle theory is based on geometrical optics, and wave optics is used to determine the effect of edges and asperities. The fine-particle theory considers the particles as ellipsoidal dipoles, randomly oriented, that cause coherent (Fresnel) scattering and noncoherent (Rayleigh) scattering.

a. Coarse-Particle Theory

Each particle is replaced by a sphere of equivalent volume with edges and surface asperities represented by dipoles distributed uniformly on the surface. Consider a collimated beam of radiation entering one of these spheres at an incident angle θ (Fig. II-4).

The angle of refraction θ_i' is given by

$$\tan \theta_i' = \sin \theta / \mathrm{Re}(m^2 - \sin^2 \theta_i)^{1/2} \qquad \text{(II-60)}$$

where Re designates the real part.

The angle of deviation φ is

$$\varphi = 2\theta - \theta_i' \qquad \text{(II-61)}$$

A wave entering the sphere travels a distance l $(=2a \cos \theta_i')$ with transmission T,

$$T = \exp(-4\pi\kappa_0 l/\lambda) \qquad \text{(II-62)}$$

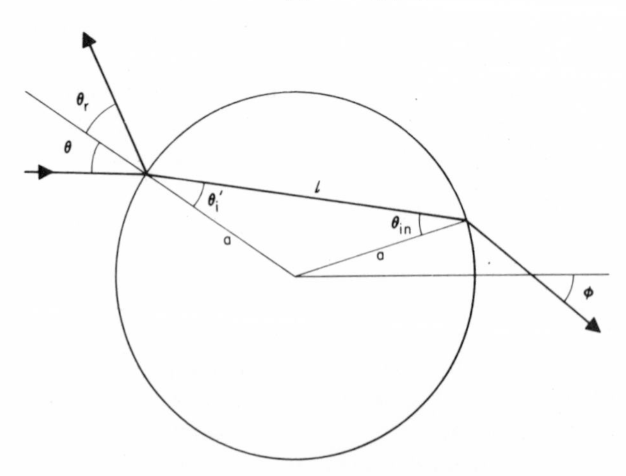

FIG. II-4 Ray tracing geometry.

If $\mathscr{I}_0$ is the unpolarized incident radiation on the sphere, then, in the angular range $d\theta$, the incident power dP_0 is

$$dP_0 = \pi a^2 \mathscr{I}_0 \sin 2\theta \, d\theta \tag{II-63}$$

and the refracted power is

$$(dP_0)_{\text{refr}} = \pi a^2 \mathscr{I}_0 T_0{}^2 T \sin 2\theta \, d\theta \tag{II-64}$$

where T_0 is the Fresnel surface transmittance, to be defined subsequently along with the surface reflectance R_0.

If we apply a weighting factor of $\cos^2 \varphi$ for the forward beam and $\sin^2 \varphi$ for the combined transverse directions, we find the forward-, transverse-, and backward-scattering cross sections due to refraction,

$$\sigma_{\text{refr,f}} = \pi a^2 \int_{\phi \leq \pi/2} T_0{}^2 T \cos^2 \varphi \sin 2\theta \, d\theta \tag{II-65}$$

$$\sigma_{\text{refr,t}} = \pi a^2 \int_{\theta=0}^{\pi/2} T_0{}^2 T \sin^2 \varphi \sin 2\theta \, d\theta \tag{II-66}$$

$$\phi_{\text{refr,b}} = \pi a^2 \int_{\phi \geq \pi/2} T_0{}^2 T \cos^2 \varphi \sin 2\theta \, d\theta \tag{II-67}$$

The externally reflected radiation cross sections are

$$\sigma_{\text{refl,f}} = \frac{\pi a^2}{8} \int_{\pi/4}^{\pi/2} R_0 \sin^2 \theta \sin 2\theta \, d\theta \tag{II-68}$$

$$\sigma_{\text{refl,t}} = \pi a^2 \int_0^{\pi/2} R_0 \left(\frac{1}{4} + \frac{1}{4} \cos^2 \theta + \frac{1}{2} \sin^2 \theta \right) \sin 2\theta \, d\theta \tag{II-69}$$

$$\sigma_{\text{refl,b}} = \pi a^3 \int_0^{\pi/4} R_0 \left(\frac{1}{8} \sin^2 \theta + \frac{1}{2} \cos^2 \theta \right) \sin 2\theta \, d\theta \tag{II-70}$$

The cross sections for multiply internally reflected radiation (intref) are

$$\sigma_{\text{intref,f}} = \frac{1}{6} \pi a^2 \int_0^{\pi/2} \frac{R_0 T_0{}^2 T^2}{1 - R_0 T} \sin 2\theta \, d\theta \tag{II-71}$$

$$\sigma_{\text{intref,t}} = \frac{4}{6} \pi a^2 \int_0^{\pi/2} \frac{R_0 T_0{}^2 T^2}{1 - R_0 T} \sin 2\theta \, d\theta \tag{II-72}$$

$$\sigma_{\text{intref,b}} = \frac{1}{6} \pi a^2 \int_0^{\pi/2} \frac{R_0 T_0{}^2 T^2}{1 - R_0 T} \sin 2\theta \, d\theta \tag{II-73}$$

The absorption cross section is the integrated sum of the factors $T_0^2 T$, R_0, and $R_0 T_0^2 T^2/(1 - R_0 T)$ subtracted from unity,

$$\sigma_{\text{abs}} = \pi a^2 \int_0^{\pi/2} \left(1 - R_0 - \frac{T_0^2 T}{1 - R_0 T}\right) \sin 2\theta \, d\theta \qquad \text{(II-74)}$$

The surface transmission and reflection factors T_0 and R_0 are functions of the Fresnel transmission and reflectance as well as the effect of edges and asperities. The effect of edges and asperities on the reflection level depends upon the ratio of the scattering by the edges and asperities to the absorption in the bulk material. Aronson and Emslie (1973) showed experimentally that very fine particles adhering to larger particles can cause a reduction in reflection level; however, Egan *et al.* (1973b) show that finer particles can also cause an increase in reflection level. The Aronson and Emslie (1975) theory partially takes into account the increase in reflection level, whereas the modified six-flux model, to be described subsequently, takes into account both effects completely. The direction of the change in brightness depends upon whether the absorption is large compared to the scattering: for highly absorptive materials, the brightness will decrease with increase in surface roughness, and, for relatively low absorptive materials, the brightness will increase with increase in surface roughness.

The Emslie and Aronson (1973) theory assumes that the edges and asperities can be represented as ellipsoidal particles. Then, using considerations of electromagnetic theory, and assuming the restricting condition that surface reflectance decreases with increased surface roughness, an expression for surface absorptance A_0 is derived,

$$A_0 = 1 - e^{-X} \qquad \text{(II-75)}$$

where

$$X = (N_e W_e + N_a W_a)/\mathscr{I}_0 \cos\theta \qquad \text{(II-76)}$$

Here N_e is the average number of edges per unit area, N_a the average number of asperities per unit area, $1/\cos\theta$ the obliquity factor, $\mathscr{I}_0$ the power density of the collimated incident beam, W_e the power absorbed by an edge, and W_a the power absorbed by an asperity.

The expression for X [Eq. (II-76)] depends upon N_e, N_a, the length of the edges, and the volume of the asperities, respectively, which are adjustable to fit the experimental data.

The Fresnel surface reflectance R_0 and transmittance T_0 are then given by

$$R_0 = |\tfrac{1}{2}(R_p + R_s)|(1 - A_0) \qquad \text{(II-77)}$$

$$T_0 = \{1 - |\tfrac{1}{2}(R_p + R_s)|\}(1 - A_0) \qquad \text{(II-78)}$$

where R_p and R_s are the complex Fresnel specular reflectivity components for radiation parallel and perpendicular to the plane of incidence. Also, for energy conservation,

$$R_0 + T_0 + A_0 = 1 \tag{II-79}$$

In order to bring the single-particle cross sections for absorption, reflection, and refraction given by Eqs. (II-65) and (II-74) into a form that applies to a powder, we must apply a radiative transfer theory. Two approaches may be used to set up the six-flux model: it may be set up (1) with two fluxes along the x direction and the remaining four fluxes along the y and z directions, or (2) with three $+x$-directed fluxes and three $-x$-directed fluxes at angles $\cos^{-1}(1/\sqrt{3})$ to the x axis. The first approach was used by Emslie and Aronson (1973) in their initially reported work on powders, and the second approach was used by Aronson and Emslie (1975) in a refined version of the theory.

The first approach will be presented here; $\mathscr{I}$ and $\mathscr{J}$ are assumed to be in the x and $-x$ directions and fluxes I_1, I_2, I_3, and I_4 to be in the transverse orthogonal directions y, $-y$, z, $-z$. Using Eq. (II-10), the individual radiative transfer equations are

$$\frac{d\mathscr{I}}{dx} = -(K + S_b + S_t)\mathscr{I} + S_b\mathscr{J} + \frac{1}{4}S_t(I_1 + I_2 + I_3 + I_4) \tag{II-80}$$

$$-\frac{d\mathscr{J}}{dx} = -(K + S_b + S_t)\mathscr{J} + S_b\mathscr{I} + \frac{1}{4}S_t(I_1 + I_2 + I_3 + I_4) \tag{II-81}$$

$$\frac{dI_1}{dy} = -(K + S_b + S_t)I_1 + S_b I_2 + \frac{1}{4}S_t(\mathscr{I} + \mathscr{J} + I_3 + I_4) \tag{II-82}$$

$$-\frac{dI_2}{dy} = -(K + S_b + S_t)I_2 + S_b I_1 + \frac{1}{4}S_t(\mathscr{I} + \mathscr{J} + I_3 + I_4) \tag{II-83}$$

$$\frac{dI_3}{dz} = -(K + S_b + S_t)I_3 + S_b I_4 + \frac{1}{4}S_t(\mathscr{I} + \mathscr{J} + I_1 + I_2) \tag{II-84}$$

$$-\frac{dI_4}{dz} = -(K + S_b + S_t)I_4 + S_b I_3 + \frac{1}{4}S_t(\mathscr{I} + \mathscr{J} + I_1 + I_2) \tag{II-85}$$

Here, K, S_b, and S_t are the absorption, backscattering, and transverse total scattering coefficients for each beam, and are related to the single-particle cross sections as follows:

$$S_t = N\sigma_\tau \tag{II-86}$$

$$S_b = N\sigma_b \tag{II-87}$$

$$K = N\sigma_{abs} \tag{II-88}$$

and

$$\sigma_\tau = \sigma_{refr,t} + \sigma_{refl,t} + \sigma_{intref,t} \tag{II-89}$$

[see Eqs. (II-66), (II-69), and (II-72)]

$$\sigma_b = \sigma_{refr,b} + \sigma_{refl,b} + \sigma_{intref,b} \tag{II-90}$$

[see Eqs. (II-67), (II-70), and (II-73)].

Note that σ_{abs} is given by Eq. (II-74). The particle density N can be related to the particle diameter d, and the volume packing fraction f by

$$N = 6f/\pi d^3 \tag{II-91}$$

Because of symmetry,

$$I_1 = I_2 = I_3 = I_4 = \tfrac{1}{4}P \tag{II-92}$$

where $P = I_1 + I_2 + I_3 + I_4$. Adding the last four radiative transfer equations [Eqs. (II-82)–(II-85)], we obtain

$$P = [S_t/(K + S_t/2)](\mathcal{I} + \mathcal{J}) \tag{II-93}$$

and Eqs. (II-80) and (II-81) may be reduced to

$$\frac{d\mathcal{I}}{dx} = (K' + S')\mathcal{I} + S'\mathcal{J} \tag{II-94}$$

$$-\frac{d\mathcal{J}}{dx} = -(K' + S')\mathcal{J} + S'\mathcal{I} \tag{II-95}$$

where

$$S' = (4KS_b + 2S_bS_t + S_t^2)/(4K + 2S_t) \tag{II-96}$$

$$K' = (4K^2 + 6KS_t)/(4K + 2S_t) \tag{II-97}$$

This is simply the Kubelka–Munk result [Eqs. (II-13) and (II-14)] except that the coefficients K and S are different.

For a coarse particulate surface with a collimated incoming beam, $\mathcal{J}$ and $\mathcal{I}$ are zero at infinite depth. Using these boundary conditions in the radiative transfer equations [(II-94) and (II-95)], we obtain the volume reflectance

$$R_v = 1 + (K'/S') - [(K'/S')^2 + (2K'/S')]^{1/2} \tag{II-98}$$

The quantity R_v is identical to R_∞ [Eq. (II-19)] given by the Kubelka–Munk theory, except for the use of the primed values for K' and S'.

The coarse-particle theory thus far does not include the effect of very high absorption by particles. This effect may be included by replacing the derivatives on the left-hand side of Eqs. (II-80)–(II-85) by finite intensity differences $\Delta\mathcal{I}$ and $\Delta\mathcal{J}$ and solving these difference equations. For a coarse highly absorbing powder composed of particles of average radius a and packing fraction f, a cube of side a_1 can be considered to contain one average particle:

$$a_1{}^3 = \tfrac{4}{3}\pi a^3/f \tag{II-99}$$

A monolayer composed of these particles would have a thickness a_1 and an average transverse area $a_1{}^2$, for each particle. The monolayer absorption α_m and reflectance ρ_m are

$$\alpha_m = \sigma_{abs}/a_1{}^2 = \sigma_{abs}/(\tfrac{4}{3}\pi a^3/f)^{2/3} \tag{II-100}$$

$$\rho_m = (\sigma_b + \tfrac{1}{2}\sigma_\tau)/a_1{}^2 = (\sigma_b + \tfrac{1}{2}\sigma_\tau)/(\tfrac{4}{3}\pi a^3/f)^{2/3} \tag{II-101}$$

where σ_{abs} is given by Eq. (II-74), σ_b by Eq. (II-90) and σ_τ by Eq. (II-89). Interparticle shadowing is not included in the foregoing expressions.

Using the expressions for α_m and ρ_m, the reflectance R_v of a semi-infinite highly absorbing stack of monolayers is

$$R_v = \frac{1 + \rho_m{}^2 - \tau_m{}^2}{2\rho_m} - \left[\left(\frac{1 + \rho_m{}^2 - \tau_m{}^2}{2\rho_m}\right)^2 - 1\right]^{1/2} \tag{II-102}$$

where the monolayer transmittance τ_m is given by

$$\tau_m = 1 - \alpha_m - \rho_m \tag{II-103}$$

For mixtures of particles, or distributed sizes, the appropriate σ_{abs}, σ_b, σ_τ are computed for each, and α_m and σ_m are the summations over all components or sizes with the corresponding packing fraction for each.

b. Fine Particles

The ray tracing approach for coarse particles breaks down when the particles, edges, and asperities are smaller than the wavelength of the incident radiation. By assuming ellipsoidal particles with a wide range of depolarization factors (by varying the ellipsoidal shapes), we can separately average the scattered and absorbed powers from both states of polarization of the incident wave. The result for the total scattering cross section σ_{sc} and the absorption cross section σ'_{abs} were given by Emslie and Aronson (1975):

$$\sigma_{sc} = \frac{4\pi^5 v^4 (2a)^6}{27} |\hat{m}^2 - 1|^2 \frac{\mathrm{Im}(\ln \hat{m}^2) + \mathrm{Im}[(\ln \hat{m}^2)/(\hat{m}^2 - 1)]}{\mathrm{Im}(\hat{m}^2)} \tag{II-104}$$

$$\sigma'_{abs} = \tfrac{2}{3}\pi^2 v (2a)^3 \,\mathrm{Im}[-\hat{m}^2(\ln \hat{m}^2)/(\hat{m}^2 - 1)] \tag{II-105}$$

(where Im designates the imaginary part, and v the frequency in wave-number units), and

$$\sigma_b{}' = \sigma_{sc}/6 \tag{II-106}$$

$$\sigma_\tau{}' = 4\sigma_{sc}/6 \tag{II-107}$$

Coherence in the scattering is taken into account by finding an average complex index $\overline{m}$ for the medium by means of a Lorentz–Lorenz type calculation. The average scattering is then computed using the relative index $\hat{m}/\overline{m}$.

Also, because of the small changes in transferred radiation across a monolayer of fine particles, a continuum form of the radiative transfer equations must be used (rather than a discrete monolayer model).

For a surface containing N particles per unit volume, the scattering coefficients are

$$S_\tau = N\sigma_\tau' \tag{II-108}$$

$$S_b = N\sigma_b' \tag{II-109}$$

and the absorption coefficient K is

$$K = N\sigma_{abs}' \tag{II-110}$$

The radiative transfer analysis for fine particles assumes that both the incident collimated flux I_0 and outgoing scattered flux J undergo a Fresnel reflection R_F caused by a Lorentz–Lorenz index of refraction of the medium $\overline{m}$, where

$$R_F = |(\overline{m} - 1)/(\overline{m} + 1)|^2 \tag{II-111}$$

and

$$\overline{m}^2 = 1 + \frac{2f\{[1 + (1/\hat{m}^2 - 1)]\ln(\hat{m}^2) - 1\}}{1 - f + [2f/(\hat{m}^2 - 1)]\{[1 + (1/\hat{m}^2 - 1)]\ln(\hat{m}^2) - 1\}} \tag{II-112}$$

Here, f is the volume fraction and $\hat{m}$ is the true index of refraction of the particle.

Under these conditions, the surface reflectance is

$$R = R_F + [(1 - R_F)^2 R_v/(1 - R_F R_v)] \tag{II-113}$$

where R_v is given by Eq. (II-98).

c. Coarse to Fine Particle Bridging

The fine-particle theory is valid when $a \ll \lambda$, and the coarse particle theory is valid when $a \gg \lambda$. Each theory gives cross sections that are too large in the nonapplicable region. The main justification for a bridging theory is the similarity of the reflection spectra in the transition region. In the transition region, a bridging formula that weighs more strongly the theory that gives the smaller cross section is

$$\sigma_{br}^{-2} = \sigma_c^{-2} + \sigma_f^{-2} \tag{II-114}$$

where

$$\sigma_c = \sigma_b + \tfrac{1}{2}\sigma_\tau \qquad \text{(II-115)}$$

$$\sigma_f = \sigma_b' + \tfrac{1}{2}\sigma_\tau' \qquad \text{(II-116)}$$

where σ_{br} is the bridged value and

$$\sigma_{absbr} = (\sigma_{br}/\sigma_0)\sigma_{abs,0} \qquad \text{(II-117)}$$

where σ_0 is the smaller of σ_c and σ_f and $\sigma_{abs,0}$ is the absorption cross section calculated by the theory that produces σ_0.

4. Modified Coarse-Particle Theory for Six-Flux Model

The six-flux model of Emslie and Aronson (1973), previously described, includes the absorption by asperities but not the scattering. The scattering of radiation by asperities is not significant in the far infrared spectral regions for which the theory was developed. However, in applying the theory to the visible, near infrared, and ultraviolet regions of the spectrum, poorer matching to observed reflectivity spectra occurs because of the stronger effect of scattering. Thus, increased scattering in the ultraviolet range would tend to raise the reflectivity; when scattering is not included, the ultraviolet reflectivity in the model must be increased by decreasing particle size, for instance. This, then raises the entire model reflectivity disproportionately producing an excessively high infrared reflectivity.

A modified six-flux model including scattering as well as asperity absorption for predicting the specular reflectance of particulate materials was developed by Egan and Hilgeman (1978a). The theory applies for particles with diameters $d \gtrsim \lambda$, and is based on a Monte Carlo approach. Thus, the Monte Carlo model is restricted to coarse particles as the region of most practical interest for remote-sensing applications. The Monte Carlo approach permits assigning probabilities for external and internal scattering by asperities of a single spherical particle, as well as for reflection and refraction. Fluxes for forward, transverse, and backward scattering are determined probabilistically for a single particle, and these are combined using the Emslie and Aronson radiative transfer theory to give the reflectance of a semi-infinite powder surface.

a. Single-Particle Theory

The model for the theory is a sphere of the average particle size with the effects of scattering and absorption described by appropriate probabilities. Rays approach a sphere from the $+Z$ direction with a uniform distribution in the XY plane. The sphere is located at the origin of the axes with radius *a*. The rays of unit intensity produce discrete interactions when they strike the surface of the sphere where each may be reflected (R), transmitted (T), absorbed (A), or scattered (S), with further combinations. The probability of an interaction may be expressed in terms of θ (the angle of an incident ray to the local surface normal at the initial or subsequent interaction), the particle size, the optical complex index of refraction ($\hat{m} = n + i\kappa_0$), a surface density of absorbers $\hat{x}$, and a surface density of scatterers $\hat{y}$. There is no azimuthal dependence in the initial interaction, but in subsequent interactions there is. Within the sphere, after transmission, the rays are attenuated according to the Beer–Lambert law.

The surface interaction probability is constrained by

$$R + T + A + S = 1 \tag{II-118}$$

where $R = R(\theta, n, \kappa_0)$, $T = T(\theta, n, \kappa_0)$, $A = A(\hat{x}, \hat{y}, \theta, n, \kappa_0)$, and $S = S(\hat{x}, \hat{y}, \theta, n, \kappa_0)$. There can be surface interactions with rays approaching from both inside and outside of the sphere for all four, R, T, A, and S (Fig. II-5). Thus, there may be R_{oo}, R_{ii}, T_{oi}, A_o, A_i, S_o, or S_i, where the subscripts i and o denote the location of the incoming ray and outgoing rays successively inside or outside of the sphere. For scattering, the single subscripts will be used to refer to the location of the outgoing ray. For absorption, since there is no

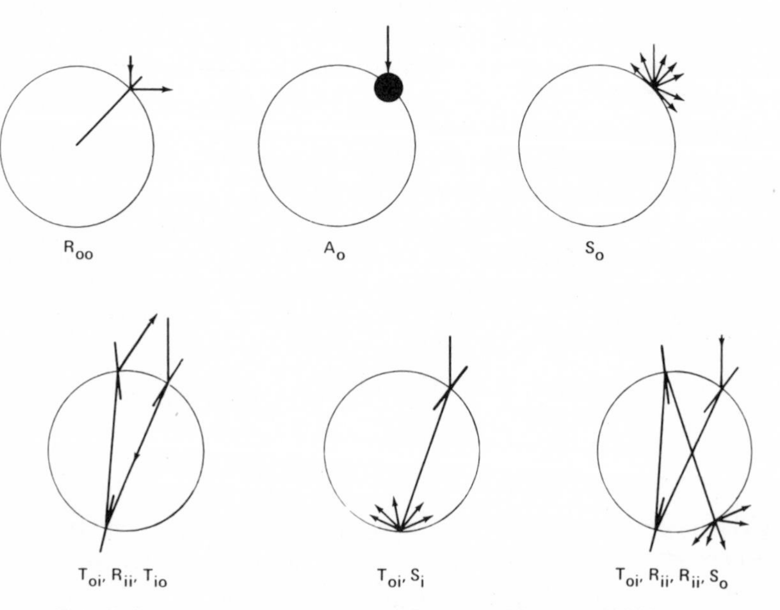

FIG. II-5 Interaction modes. [From Egan and Hilgeman (1978a).]

outgoing ray, single subscripts are used to refer to the location of the incoming ray.

The reflectivity R is given by

$$R = |\tfrac{1}{2}(R_s + R_p)| \tag{II-119}$$

where R_s and R_p are the complex Fresnel reflectivities (i.e., an average is taken over the polarization components).

The output from the interactions for those rays not absorbed on the surface or attenuated from too long an internal path will be rays emerging in various directions. These rays will be divided into forward, backward, and transverse directions defined in terms of spherical coordinates relative to the Z axis.

For interactions which leave the ray inside the sphere, the location of the next interaction point must be calculated. For interactions which leave the ray outside of the sphere, the attenuated intensity must be added to the running summations for the various directions.

R_{oo}: This interaction is simply described (Fig. II-4) by Fresnel's reflection equation; here θ varies between 0 and $\pi/2$ and θ_r goes from 0 to $\pi/2$. The angle of deviation of the outgoing ray relative to the Z axis is ϕ $(=2\theta)$; if $\phi \leq \pi/2$, backscatter b_s occurs (proportional to $\cos^2 \phi$). If $\phi > \pi/2$, forward scatter f_s occurs (proportional to $\cos^2 \phi$), and for all angles $(0 \leq \phi \leq \pi)$, the transverse scatter t_s is proportional to $\sin^2 \phi$.

R_{ii}: Inside reflection may occur at any point within the sphere. The probability of reflectivity is also given by Eq. (II-119) with the quantities R_s and R_p redefined appropriately for inside of the sphere. In order to calculate the location of the next interaction point, a local orthogonal coordinate system is introduced containing Z' and X' axes with the third axis being the radius vector r of the sphere. The $X'Z'$ tangent plane has axis Z' lying in the Zr plane (Fig. II-6). The emergent ray (at angle θ to the particle radius) makes an angle ϵ to Z' to strike the sphere at a point located by spherical-triangle angles β and Q*. The angle of deviation Δ of the reflected ray from the Z axis is another arc of this spherical triangle.

In the spherical triangle (Fig. II-6), it can be shown that

$$\cos \beta = \sin \Delta \sin 2\theta \cos \epsilon - \cos \Delta \cos \theta \tag{II-120}$$

$$\sin Q = -\sin \Delta \sin \epsilon / \sin \beta \tag{II-121}$$

$$\theta_{in} = \theta \tag{II-122}$$

Also the internally reflected ray will be attenuated as it traverses the path l by a factor $\exp(-4\pi\kappa_0 l/\lambda)$, where $l = 2a \cos \theta$. With β, Q, θ, and the attenuation

* Here β and ϵ are unrelated to the previous usage in this chapter.

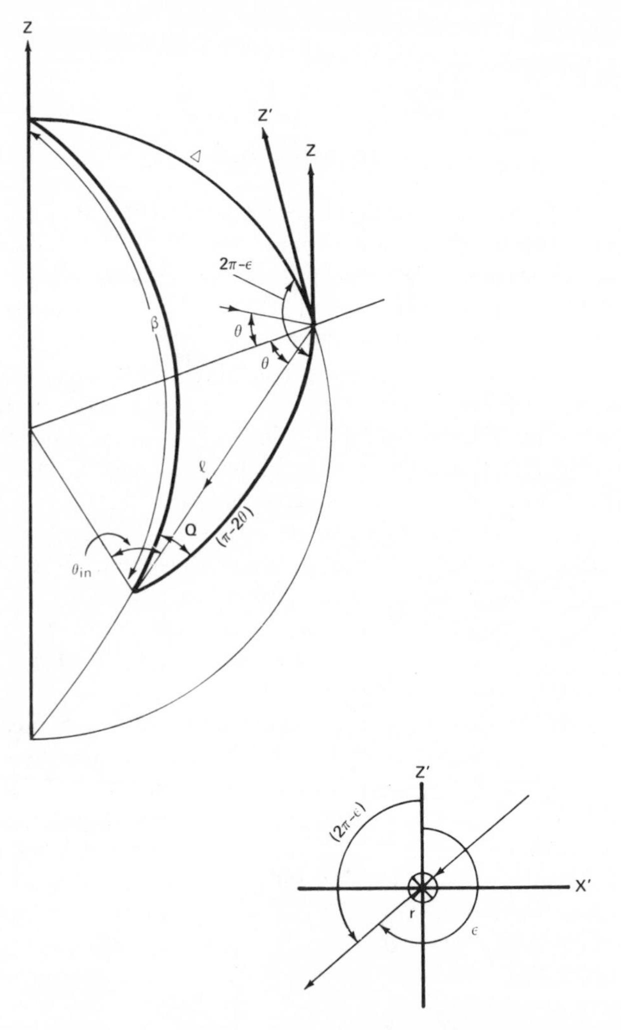

FIG. II-6 Inside reflection (R_{ii}) geometry. [From Egan and Hilgeman (1978a).]

of the reflected ray, all the parameters of the ray coming to the next interaction point are defined.

T_{oi}: External transmission (refraction inward at the spherical surface, Fig. II-4) is given by Snell's law,

$$\tan \theta_{\text{refr}} = \sin \theta / \text{Re}(\hat{m}^2 - \sin^2 \theta_i')^{1/2} \tag{II-123}$$

The internally transmitted ray will be attenuated over the length l within the sphere. In order to specify the next interaction point, we need the elevation angle $\theta_{\text{in}} = \theta_i'$, and the angle β (the spherical angle between the Z axis and the interaction point), given by

$$\beta = \Delta + \pi - 2\theta_i' \qquad \text{or} \qquad = 2\theta_i' - \Delta \tag{II-124}$$

such that $\beta < \pi$ and Δ is the spherical angle between the Z axis and the incident point.

T_{io}: This interaction is shown in Fig. II-7. Snell's law relates θ_0 to θ, but the index of refraction is $1/\hat{m}$ going out of the particle (Fig. II-7a). The angle of the incoming ray to the Z' axis is Q (Fig. II-7b). Also required is the angle ϕ of the emerging ray to the Z axis,

$$\cos\phi = \sin\theta_0 \cos\Delta - \sin\theta_0 \sin\Delta \cos Q \tag{II-125}$$

obtained from the spherical-triangle relationships (Fig. II-7c).

A_o, A_i: Absorption, for our purposes, is considered as taking place on the surface of the sphere in a probabilistic manner depending upon the surface density of "absorbers" (edges or asperities). A ray considered absorbed disappears entirely.

S_i: Scattering inside is assumed to be symmetrical in azimuth ϵ, and proportional to $\cos\theta$ with $0 \le \theta \le \pi/2$; the projection of the scattered ray on the $X'Z'$ plane is p; Z, Z', and r are coplanar; Z' is orthogonal to r and X'; Z', X', and p are coplanar; and the scattered ray, r, and p are also coplanar (Fig. II-8a, b). The next interaction point of the scattered ray on the sphere is defined by θ, Q, and β. In the spherical triangle in Fig. II-8c,

$$\cos\beta = -\cos\Delta\cos 2\theta + \sin\Delta\sin 2\theta\cos\epsilon \tag{II-126}$$

$$\sin(2\pi - Q) = (\sin\epsilon/\sin\beta)\sin\Delta \tag{II-127}$$

and

$$\theta_{\text{new}} = \theta. \tag{II-128}$$

The internally scattered ray will be attenuated by $\exp(-4\pi\kappa_0 l/\lambda)$, where $l = 2a\cos\theta$, before striking the sphere internally again, that interaction being governed by the appropriate interaction probability for R, T, A, or S.

S_o: Beyond the probability selections, as in S_i, there is also the determination of the weighted distribution of the forward, backward, and transverse energies. Consider the scattering to be Lambertian and the surface normal to lie in the YZ plane (Fig. II-9). The symmetry of the solution allows the X and Y axes to be so defined. The flux along the Z axis is required ($+Z$ for backscatter, $-Z$ for forward scatter). This flux is obtained by integrating over the locus of the scattered radiation as a function of direction given by

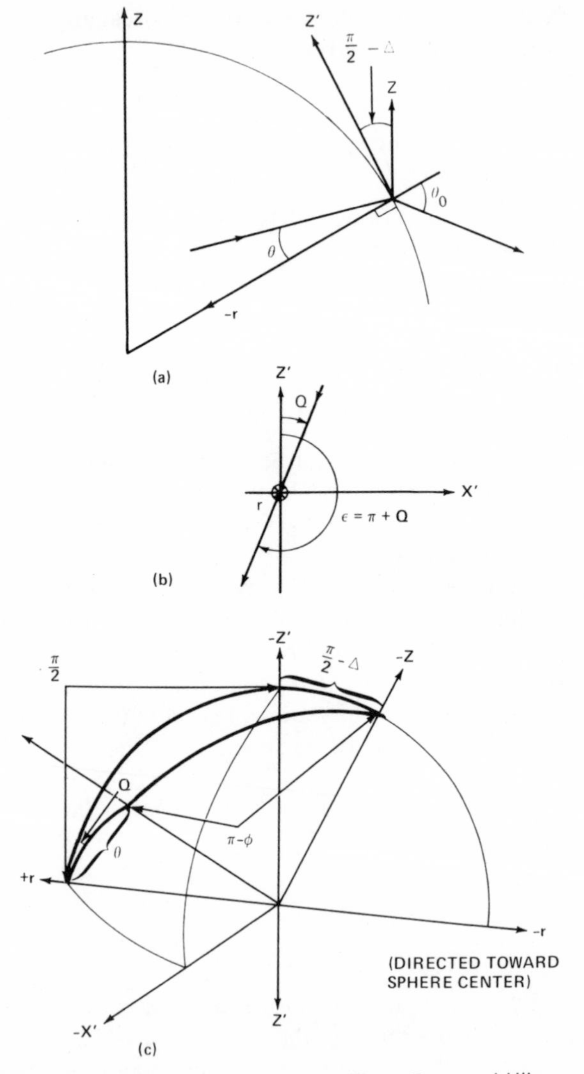

FIG. II-7 Inside transmission (T_{io}) geometry. [From Egan and Hilgeman (1978a).]

angles β, γ, δ, and ξ. The backward, forward, and transverse fluxes are given by

$$W_b = \tfrac{1}{8}(1 + \cos\beta)^2 \tag{II-129}$$

$$W_f = \tfrac{1}{8}(1 - \cos\beta)^2 \tag{II-130}$$

$$W_t = 1 - W_b - W_f = \tfrac{1}{4}(3 - \cos^2\beta) \tag{II-131}$$

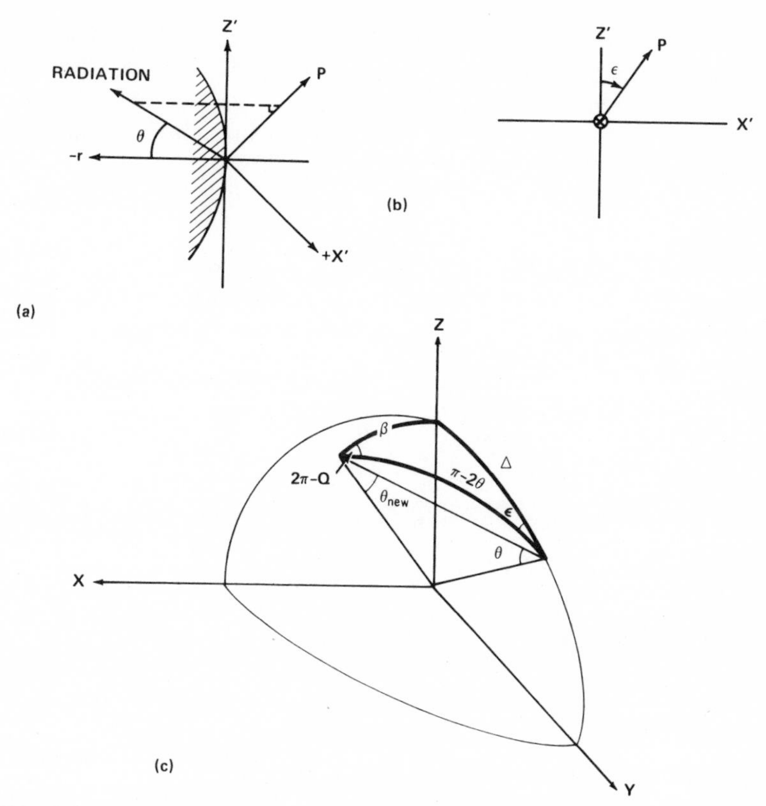

FIG. II-8 Inside scattering (S_i) geometry. [From Egan and Hilgeman (1978a).]

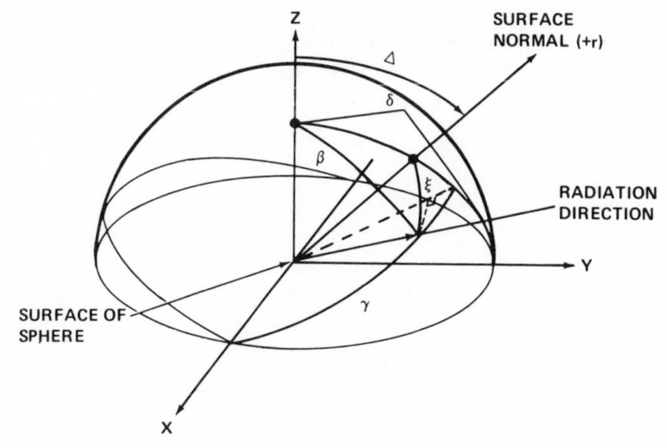

FIG. II-9 Outside scattering (S_o) geometry. [From Egan and Hilgeman (1978a).]

Another parameter, which is designated SI, has been added to the theory; this is the fraction of total scattered radiation which is scattered internally. It may vary between 0 and 1.

Scattering and absorption by asperities are brought into the theory through the factors S and A:

$$S = 1 - \exp(-\hat{y}/\cos\theta) \tag{II-132}$$

$$A = 1 - \exp(-\hat{x}/\cos\theta) \tag{II-133}$$

The relative effect of R, T, S, and A is linked to the random method for interaction selection. A six-digit random number (RN) between 0 and 1 directs the selection.

If

$$\exp(-\hat{y}/\cos\theta) < \mathrm{RN} \tag{II-134}$$

then a scatter S occurs. If

$$\exp(-\hat{y}/\cos\theta) + \exp(-\hat{x}/\cos\theta) - 1 < \mathrm{RN} < \exp(-\hat{y}/\cos\theta) \tag{II-135}$$

then absorption A occurs. If

$$[\exp(-\hat{y}/\cos\theta) + \exp(-\hat{x}/\cos\theta) - 1)][\tfrac{1}{2}(R_p + R_s)] < \mathrm{RN} < \exp(-\hat{y}/\cos\theta) + \exp(-\hat{x}/\cos\theta) - 1 \tag{II-136}$$

then reflection R occurs. Otherwise, there is transmission T. The R, T, A, and S probabilities thus satisfy Eq. (II-118).

The selection of an exponential scattering and absorption dependence on $\hat{x}$ and $\hat{y}$ simplifies the random number selection process. The actual values for $\hat{x}$ and $\hat{y}$ are determined empirically, as well as that for SI.

The results of a Monte Carlo calculation yield the forward F_x, backward B_x, and transverse T_x scattering cross sections; subtracting these from unity yields the absorption cross section. In analogy to Eqs. (II-86) through (II-91), we define the forward F_s, backward B_s, and transverse T_s scattering coefficients. Thus,

$$1 - F_s/N - B_s/N - T_s/N = K_s/N \tag{II-137}$$

where N is given by Eq. (II-91)

b. Radiative Transfer

The scattering S' and absorption K' for a multiply scattering layer with an incident parallel beam of radiation, in analogy to Eqs. (II-96) and (II-97), are

$$S' = (4K_s B_s + 2B_s T_s + T_s^2)/(4K_s + 2T_s) \tag{II-138}$$

$$K' = (4K_s^2 + 6K_s T_s)/(4K_s + 2T_s) \tag{II-139}$$

where B_s, T_s, and K_s are identified with the quantities S_b, S_t, and K of the Emslie and Aronson theory and are treated accordingly in their radiative transfer theory [Eqs. (II-96), (II-97), (II-98)].

5. Modified Dispersion Model

The Kramers–Kronig dispersion model discussed in Chapter I requires a fundamental restructuring to include the effects of scattering (Young, 1977). This modification is beyond the scope of the present text. However, the effect of scattering on the measured optical complex index of refraction will be indicated. One of the methods for transforming the phase and amplitude of the reflected amplitude into the actual complex optical constants is with a Smith diagram (Fig. II-10). Figure II-10 is a polar plot illustrating the use of the Smith diagram. Here, the amplitude of the reflectivity r and the phase shift ϕ [Eq. (I-44)] are plotted in polar form. Then the corresponding real and imaginary portions of the optical complex index of refraction derived from Eq. (I-41) may be conveniently read from the plot.

Scattering will lower the specularly reflected radiation, and also reduce the phase shift. In the example shown in Fig. II-10, the arrow indicates that a material with a true absorption coefficient of 0.8 and real index of 1.5 in the

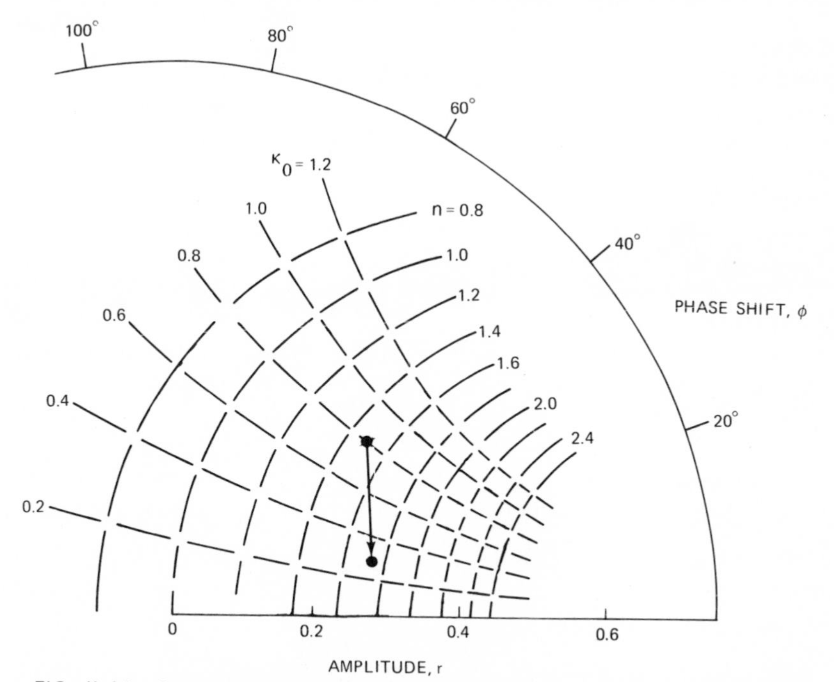

FIG. II-10 Smith diagram showing amplitude and phase of complex reflection coefficient and the effect of scattering for normal incidence.

presence of scattering could appear to have an absorption coefficient of 0.3 and a real index of 1.7. The exact amount of the shift cannot be calculated without a knowledge of the details of the scattering. The effect of the scattering on n and κ_0 depends upon where the complex reflection coefficient falls on the Smith diagram.

C. ATMOSPHERIC SCATTERING

1. Scattering by Spheres (Single Scattering)

The theoretical models that have been considered thus far in this chapter have their primary application to solids, and are based on approximate solutions for absorption and scattering by a single particle. If diffraction phenomena are included, a more exact representation of the scattering by small particles is obtained. One approach to the analysis of scattering and absorption is to assume that the particles are spherical. This simplifies the analysis because solutions of the wave equation [Eq. (I-11)] are available in terms of standard mathematical functions.

Thus, the scalar wave equation

$$\nabla\Psi + k^2\hat{m}^2\Psi = 0 \tag{II-140}$$

is separable in polar coordinates (r, θ, φ), with solutions

$$\Psi_{ln} = \left.\begin{matrix}\cos l\varphi\\ \sin l\,\varphi\end{matrix}\right\} P_n{}^l(\cos\theta) Z_n(\hat{m}kr) \tag{II-141}$$

where n, l, are integers $n \geq l \geq 0$, $\hat{m}$ is the optical complex index of refraction, k the propagation constant, $P_n{}^l$ the associated Legendre polynomial, and $Z_n(\hat{m}kr)$ the spherical Bessel function.

Using the boundary conditions for a plane wave scattered by a homogenous sphere in a vacuum, Mie (1908) derived expressions for scattering (C_{sca}) and absorption (C_{abs}) cross sections for a sphere. The sum of the scattering and absorption cross sections is the extinction cross section C_{ext}

$$C_{ext} = C_{sca} + C_{abs} \tag{II-142}$$

Frequently, efficiency factors for extinction Q_{ext}, absorption Q_{abs}, and scattering Q_{sca} are used, which are related to the geometrical cross section of a sphere of radius a $(=\pi a^2)$

$$Q_{ext} = C_{ext}/\pi a^2 \tag{II-143}$$

$$Q_{sca} = C_{sca}/\pi a^2 \tag{II-144}$$

$$Q_{abs} = C_{abs}/\pi a^2 \tag{II-145}$$

Thus, the "efficiency" of any of the processes is related to what a sphere would produce when a plane parallel wave is incident upon it.

The Mie solutions for a single particle may be expressed analytically (van de Hulst, 1957), but for practical use, either tables of Mie functions or computer programs are available (see Appendix C). These generally list efficiency factors as a function of the optical complex index of refraction $\hat{m}$, and a dimensionless size parameter x,

$$x = 2\pi a/\lambda \qquad \text{(II-146)}$$

Here, a is the radius of the sphere and λ the wavelength of the incident radiation.

If the intensity of the incident radiation is I_0, the parts absorbed and scattered are given by $Q_{\text{abs}}\pi a^2 I_0$ and $Q_{\text{sca}}\pi a^2 I_0$, respectively. If there are N particles per unit volume, these factors will be multiplied by this quantity.

The Mie theory considers each of the scattering spheres as independent, without interparticulate scattering. This can introduce a considerable error in scattering calculations involving the earth's atmosphere where multiple scattering occurs. The theory is applicable only when the particles are far apart.

The scattered radiation from a particle may also be presented as a function of the scattering angle; this angle is the direction of the scattered radiation

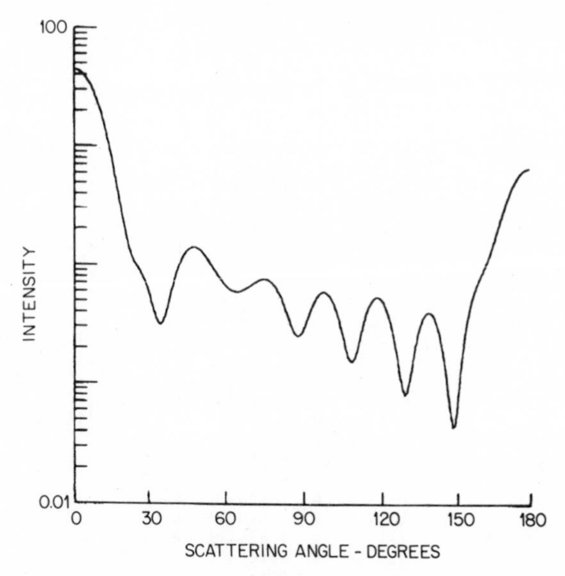

FIG II-11 Intensity of the scattered radiation corresponding to the scattering angle. Size parameter = 8.48, n = 1.902, κ_0 = 0.001395, normalization factor = $4.0/Q_s \times$ (size parameter)^2.

relative to the incident direction. The functional dependence on $\hat{m}$, x, and scattering angle is illustrated in an example for $\hat{m} = 1.902 - i0.001395$ and a size parameter $x = 8.48$ (Fig. II-11 and Table II-2).

In discussing Mie scattering, both single and multiple, a very compact notation is achieved using matrices. This description of electromagnetic radiation (including polarization) is by means of a Stokes vector **I** (Chandrasekhar, 1960)

$$\mathbf{I} = \{I, Q, U, V\} \tag{II-147}$$

where I gives amplitude, Q gives linear polarization, U yields plane of polarization, V yields circular polarization.

TABLE II-2 Intensity of the Scattered Radiation Corresponding to the Scattering Angle[a]

THETA	THETA	THETA + 1	THETA + 2	THETA + 3	THETA + 4
0.00	4.40949E 1	4.37869E 1	4.28742E 1	4.13910E 1	3.93919E 1
5.00	3.69493E 1	3.41494E 1	3.10880E 1	2.78653E 1	2.45813E 1
10.00	2.13309E 1	1.81998E 1	1.52611E 1	1.25728E 1	1.01762E 1
15.00	8.09562E 0	6.33868E 0	4.89811E 0	3.75378E 0	2.87553E 0
20.00	2.22624E 0	1.76504E 0	1.45031E 0	1.24244E 0	1.10600E 0
25.00	1.01141E 0	9.35907E -1	8.63853E -1	7.86487E -1	7.01136E -1
30.00	6.10047E -1	5.18978E -1	4.35689E -1	3.68479E -1	3.24888E -1
35.00	3.10659E -1	3.29029E -1	3.80371E -1	4.62184E -1	5.69391E -1
40.00	6.94883E -1	8.30235E -1	9.66499E -1	1.09500E 0	1.20802E 0
45.00	1.29942E 0	1.36495E 0	1.40247E 0	1.41192E 0	1.39510E 0
50.00	1.35535E 0	1.29707E 0	1.22530E 0	1.14520E 0	1.06166E 0
55.00	9.79016E -1	9.00758E -1	8.29500E -1	7.66956E -1	7.14040E -1
60.00	6.71019E -1	6.37694E -1	6.13576E -1	5.98038E -1	5.90414E -1
65.00	5.90046E -1	5.96273E -1	6.08374E -1	6.25486E -1	6.46511E -1
70.00	6.70049E -1	6.94361E -1	7.17397E -1	7.36877E -1	7.50440E -1
75.00	7.55841E -1	7.51176E -1	7.35114E -1	7.07104E -1	6.67526E -1
80.00	6.17767E -1	5.60202E -1	4.98066E -1	4.35236E -1	3.75922E -1
85.00	3.24304E -1	2.84150E -1	2.58439E -1	2.49044E -1	2.56510E -1
90.00	2.79936E -1	3.17002E -1	3.64131E -1	4.16772E -1	4.69798E -1
95.00	5.17963E -1	5.56385E -1	5.81014E -1	5.89027E -1	5.79115E -1
100.00	5.51633E -1	5.08595E -1	4.53505E -1	3.91050E -1	3.26666E -1
105.00	2.66036E -1	2.14550E -1	1.76790E -1	1.56088E -1	1.54195E -1
110.00	1.71104E -1	2.05034E -1	2.52590E -1	3.09074E -1	3.68924E -1
115.00	4.26237E -1	4.75330E -1	5.11278E -1	5.30381E -1	5.30522E -1
120.00	5.11363E -1	4.74380E -1	4.22720E -1	3.60896E -1	2.94358E -1
125.00	2.28966E -1	1.70431E -1	1.23765E -1	9.28094E -2	7.98645E -2
130.00	8.54798E -2	1.08409E -1	1.45742E -1	1.93194E -1	2.45537E -1
135.00	2.97109E -1	3.42380E -1	3.76496E -1	3.95752E -1	3.97967E -1
140.00	3.82694E -1	3.51271E -1	3.06690E -1	2.53311E -1	1.96438E -1
145.00	1.41804E -1	9.50228E -2	6.10549E -2	4.37435E -2	4.54752E -2
150.00	6.69972E -2	1.07414E -1	1.64365E -1	2.34375E -1	3.13337E -1
155.00	3.97081E -1	4.81978E -1	5.65510E -1	6.46741E -1	7.26641E -1
160.00	8.08218E -1	8.96424E -1	9.97845E -1	1.12018E 0	1.27154E 0
165.00	1.45965E 0	1.69098E 0	1.96993E 0	2.29811E 0	2.67379E 0
170.00	3.09158E 0	3.54245E 0	4.01395E 0	4.49080E 0	4.95565E 0
175.00	5.39015E 0	5.77604E 0	6.09634E 0	6.33648E 0	6.48526E 0
180.00	6.53566E 0	0.00000E 0	0.00000E 0	0.00000E 0	0.00000E 0

[a] $n = 1.902$, $\kappa_0 = 0.001395$, $x = 8.48$, $Q_e = 2.30284$, $Q_s = 2.22321$, $Q_a = 0.0796313$, normalization factor $= 4/Q_s x^2$.

The scattered radiation at a distance r ($\gg \lambda$) in a volume element is

$$K_{\text{sca}} \mathbf{P} \mathbf{I}_0 \, dv/4\pi r^2 \tag{II-148}$$

where $\mathbf{I}_0$ is the Stokes vector of the incident radiation, K_{sca} the scattering coefficient (cm^{-1}), $\mathbf{P}$ the phase matrix (4 columns, 4 rows), and

$$K_{\text{sca}} = C_{\text{sca}}/dv \tag{II-149}$$

The operator $\mathbf{P}$ produces the polarization and angular distribution of the scattered radiation for any incident radiation polarization. $\mathbf{P}$ and the quantity K_{sca} describe the radiation scattered by a small volume element. In addition, for multiple-scattering analyses, the single-scattering albedo $\tilde{\omega}$ must be known. This is the fraction of the total beam extinction that is accounted for by scattering.

Thus,

$$\tilde{\omega} = Q_{\text{sca}}/Q_{\text{ext}} = C_{\text{sca}}/C_{\text{ext}} = K_{\text{sca}}/K_{\text{ext}} \tag{II-150}$$

When P is only a function of scattering angle α_0 with at most six independent parameters,

$$\mathbf{P}(\alpha_0) = \begin{bmatrix} P_{11} & P_{21} & 0 & 0 \\ P_{21} & P_{22} & 0 & 0 \\ 0 & 0 & P_{33} & -P_{43} \\ 0 & 0 & P_{43} & P_{44} \end{bmatrix} \tag{II-151}$$

For Mie scattering, the phase matrix is given by van de Hulst (1957) as $\mathbf{P}(\alpha_0)$,

$$\mathbf{P}(\alpha_0) = \begin{bmatrix} \frac{1}{2}(\hat{S}_1\hat{S}_1{}^* + \hat{S}_2\hat{S}_2{}^*) & \frac{1}{2}(\hat{S}_1\hat{S}_1{}^* - \hat{S}_2\hat{S}_2{}^*) & 0 & 0 \\ \frac{1}{2}(\hat{S}_1\hat{S}_1{}^* - \hat{S}_2\hat{S}_2{}^*) & \frac{1}{2}(\hat{S}_1\hat{S}_1{}^* + \hat{S}_2\hat{S}_2{}^*) & 0 & 0 \\ 0 & 0 & \frac{1}{2}(\hat{S}_1\hat{S}_2{}^* + \hat{S}_2\hat{S}_1{}^*) & \frac{1}{2}(\hat{S}_1\hat{S}_2{}^* - \hat{S}_2\hat{S}_1{}^*) \\ 0 & 0 & -\frac{1}{2}(\hat{S}_1\hat{S}_2{}^* - \hat{S}_2\hat{S}_1{}^*) & \frac{1}{2}(\hat{S}_1\hat{S}_2{}^* + \hat{S}_2\hat{S}_1{}^*) \end{bmatrix} \tag{II-152}$$

The asterisk represents the complex conjugate, where

$$\hat{S}_1 = \sum_{n=1}^{\infty} \frac{2n+1}{n(n+1)} \hat{a}_n \pi_n + \hat{b}_n \tau_n \tag{II-153}$$

$$\hat{S}_2 = \sum_{n=1}^{\infty} \frac{2n+1}{n(n+1)} \hat{b}_n \pi_n + \hat{a}_n \tau_n \tag{II-154}$$

and π_n and τ_n are related to the Legendre polynomials; as a function of the scattering angle α_0,

$$\pi_1(\alpha_0) = 1 \tag{II-155}$$

$$\pi_2(\alpha_0) = 3 \cos \alpha_0 \tag{II-156}$$

$$\tau_1 = \cos \alpha_0 \tag{II-157}$$

$$\tau_2 = 3 \cos 2\alpha_0 \tag{II-158}$$

The calculation of the coefficients $\hat{a}_n$ and $\hat{b}_n$ are the basis of the Mie scattering expression, and are calculated by van de Hulst (1957), using spherical Bessel functions.

The scattering and extinction efficiency factors are then given as

$$Q_{\text{sca}} = \frac{2}{x^2} \sum_{n=1}^{\infty} (2n + 1)(\hat{a}_n \hat{a}_n{}^* + \hat{b}_n \hat{b}_n{}^*) \tag{II-159}$$

$$Q_{\text{ext}} = \frac{2}{x^2} \sum_{n=1}^{\infty} (2n + 1) \,\text{Re}(\hat{a}_n + \hat{b}_n) \tag{II-160}$$

If a size distribution $n(r)$ is imposed upon the particles (as occurs in nature), the phase matrix elements $P^{ij}(\alpha_0)$ are given by

$$P^{ij}(\alpha_0) = \int_{r_1}^{r_2} P^{ij}(\alpha_0, r) n(r) \, dr \tag{II-161}$$

where $P^{ij}(\alpha_0, r)$ is the matrix element for a particular radius r, normalized to the total number of particles per unit volume in the incremental radius dr, between the minimum radius r_1 and maximum radius r_2.

Various parameters are used to specify the particle size distribution such as effective radius, mean radius for scattering, effective variance, and skewness. Hansen and Travis (1974) give a good discussion of these.

2. General Radiative Transfer Theory

The equation of transfer is

$$-\mu \frac{d\mathscr{I}(\theta, \varphi)}{d\tau} = \mathscr{I}(\theta, \varphi) - \mathscr{J}(\theta, \varphi) \tag{II-162}$$

where $\mathscr{I}$ is the incident radiation in the direction of increasing optical depth, τ the optical depth [see Eq. (II-175) for definition], and

$$\mathscr{J}(\theta, \varphi) = \frac{1}{4\pi} \int_0^{\pi} \int_0^{2\pi} p(\theta, \varphi; \theta_s, \varphi_s) \mathscr{I}(\theta_s, \varphi_s) \sin \theta_s \, d\theta_s \, d\varphi_s \tag{II-163}$$

Here, $p(\theta, \varphi; \theta_s, \varphi_s)$ is known as the phase function, which relates the scattering intensity in the direction (θ, φ) to the radiation in the direction (θ_s, φ_s). When one considers the Stokes parameters [Eq. (II-147)], then the above equation becomes four independent equations, one for each component of the Stokes vector. The phase function p is then given by Eq. (II-161). For the remainder of this section, only the single unpolarized component of radiation will be discussed.

For plane parallel layers which occur in many applications, the general equation of transfer, expressed in terms of μ ($=\cos\theta$), is

$$-\mu\frac{d\mathscr{I}(\tau,\mu,\varphi)}{d\tau} = \mathscr{I}(\tau,\mu,\varphi) - \frac{1}{4\pi}\int_{-1}^{+1}\int_{0}^{2\pi} p(\mu,\varphi;\mu_s,\varphi_s)\mathscr{I}(\tau,\mu_s,\varphi_s)\,d\mu_s\,d\varphi_s \tag{II-164}$$

where θ is the angle to the inward surface normal.

The phase function integrated over the total 4π solid angle is the albedo for single scatter $\tilde{\omega}$ and represents the fractional part of radiation lost by scattering,

$$\int p(\text{cps}\ \theta)\frac{d\Omega}{4\pi} = \tilde{\omega} \leq 1 \tag{II-165}$$

The fraction $1 - \tilde{\omega}$ is the radiation that is absorbed. The albedo for single scatter can be expressed as a function of the scattering and absorption coefficients σ and α, respectively, as

$$\tilde{\omega}_0 = \sigma/(\sigma + \alpha) \tag{II-166}$$

For isotropic materials,

$$p(\cos\theta) = \tilde{\omega} = \sigma/(\sigma + \alpha) \tag{II-167}$$

For nonisotropic materials, the phase function may be expressed as a Legendre polynomial,

$$p(\cos\theta) = \sum_{l=0}^{\infty}\omega_l P_l(\cos\theta) \tag{II-168}$$

For isotropic scattering materials,

$$-\frac{\mu\,d\mathscr{I}(\tau,\mu)}{d\tau} = \mathscr{I}(\tau,\mu) - \frac{1}{2}\tilde{\omega}\int_{-1}^{+1}\mathscr{I}(\tau,\mu_s)\,d\mu_s \tag{II-169}$$

This integrodifferential equation may be expressed more conveniently in terms of Gaussian quadratures, the number of integration points being the number of directions for which a solution is required.

Thus,

$$-\mu_k \frac{dI_k}{d\tau} = I_k - \frac{1}{2}\tilde{\omega} \sum_{j=-n}^{+n} a_j I_j \tag{II-170}$$

where

$$a_j = \frac{1}{P_n'(\mu_j)} \int_{-1}^{+1} \frac{P_n(\mu)}{\mu - \mu_j} \, d\mu \tag{II-171}$$

and μ_j is one zero of the Legendre polynomial $P_n(\mu)$.

The solutions are given in terms of the H integrals by letting $n \rightarrow \infty$. Here,

$$H(\mu) = 1 + \mu H(\mu) \int_0^1 \frac{\psi(\mu_s)}{(\mu + \mu_s)} H(\mu_s) \, d\mu_s \tag{II-172}$$

where the characteristic function ψ is an even polynomial in μ, which depends upon the nature of the scattering, and the moments α_n of $H(\mu)$ are given by

$$\alpha_n = \int_0^1 \mu^n H(\mu) \, d\mu \tag{II-173}$$

Tables of the H and α integrals are listed by Chandrasekhar (1960).

An example of the solution for total reflectance for radiation incident in direction μ_i in an isotropic scatterer has been given by Giovanelli (1963).

$$R(\mu_i) = 1 - H(\mu_i)(1 - \tilde{\omega})^{1/2} \tag{II-174}$$

In general, the development of solutions by this technique is quite involved.

3. Multiple Scattering

Mie scattering is applicable to randomly distributed spheres, separated by distances large compared to the wavelength of the incident radiation. Under these conditions, the total scattered energy is the sum of the individually scattered energies. This applies to metallic or dielectric suspensions, atmospheric dust, the interstellar medium, the solar corona, and some water clouds.

However, when scatterers are more closely spaced, as the molecules and aerosols of the earth's atmosphere, the interrelationship between the scattering centers must be considered. This is achieved with a multiple-scattering theory. One frequently used approach is the doubling or adding method: If the reflection and transmission are known for each of two layers, a combination of these two layers will produce a transmission and reflection that is computed by considering the multiple reflections between these two layers. By choosing identical layers, the overall transmission and reflection for a thick homogeneous layer (such as the earth's atmosphere) can be built up in a geometric manner (doubling).

The absorption of planetary atmospheres is generally given in terms of an optical depth τ:

$$\tau = \int_h^\infty \eta_{\mathrm{ext}} \rho \, dh' \qquad \text{(II-175)}$$

where η_{ext} is the extinction coefficient per unit mass, ρ the atmospheric density as function of height h', and τ the optical depth ($=0$ at the top and τ_0 at the bottom).

However, for the following derivation, it is necessary to separate the effects of scattering from absorption. This requires the use of an optical depth for absorption ζ, defined as

$$\zeta = \int_h^\infty \eta_{\mathrm{abs}} \rho \, dh' \qquad \text{(II-176)}$$

where η_{abs} is the optical absorption coefficient per unit mass, ρ and h' are as in Eq. (II-175), and ζ is the optical depth as a result of absorption alone ($=0$ at the top and ζ_0 at the bottom). The geometry is shown in Fig. II-12.

Let ζ_a and ζ_b be the optical thicknesses of the top and bottom homogeneous layers a and b that are to be added (Fig. II-13). The subscript a refers to the upper layer and b to the lower layer; the indices n in the figure refer to the number of times plus one that an upward-going photon crosses the middle boundary; I_0 designates the incident radiation, R the reflected radiation, and T that transmitted. At the center of the double layer, the radiation intensities up and down are designated $\mathbf{U}$ and $\mathbf{D}$. Now, the net diffuse reflectance $\mathbf{R}(\zeta_a + \zeta_b)$ and net diffuse transmission $\mathbf{T}(\zeta_a + \zeta_b)$ can be inferred from the figure as

$$\mathbf{R}(\zeta_a + \zeta_b) = \mathbf{R}_a + \mathbf{U} \exp(-\zeta_a/\mu_s) + \mathbf{T}_a \mathbf{U} \qquad \text{(II-177)}$$

$$\mathbf{T}(\zeta_a + \zeta_b) = \mathbf{D} \exp(-\zeta_b/\mu_s) + \mathbf{T}_b \exp(-\zeta_a/\mu_i) + \mathbf{T}_b \mathbf{D} \qquad \text{(II-178)}$$

where

$$\mathbf{D} = \mathbf{T}_a + \mathbf{S} \exp(-\zeta_a/\mu_i) + \mathbf{S}\mathbf{T}_a \qquad \text{(II-179)}$$

$$\mathbf{U} = \mathbf{R}_b \exp(-\zeta_a/\mu_i) + \mathbf{R}_b \mathbf{D} \qquad \text{(II-180)}$$

$$\mathbf{S} = \sum_{m=1}^{\infty} \mathbf{Q}_m \qquad \text{(II-181)}$$

$$\mathbf{Q}_{m+1} = \mathbf{Q}_1 \mathbf{Q}_m \qquad \text{(II-182)}$$

$$\mathbf{Q}_1 = \mathbf{R}_a \mathbf{R}_b \qquad \text{(II-183)}$$

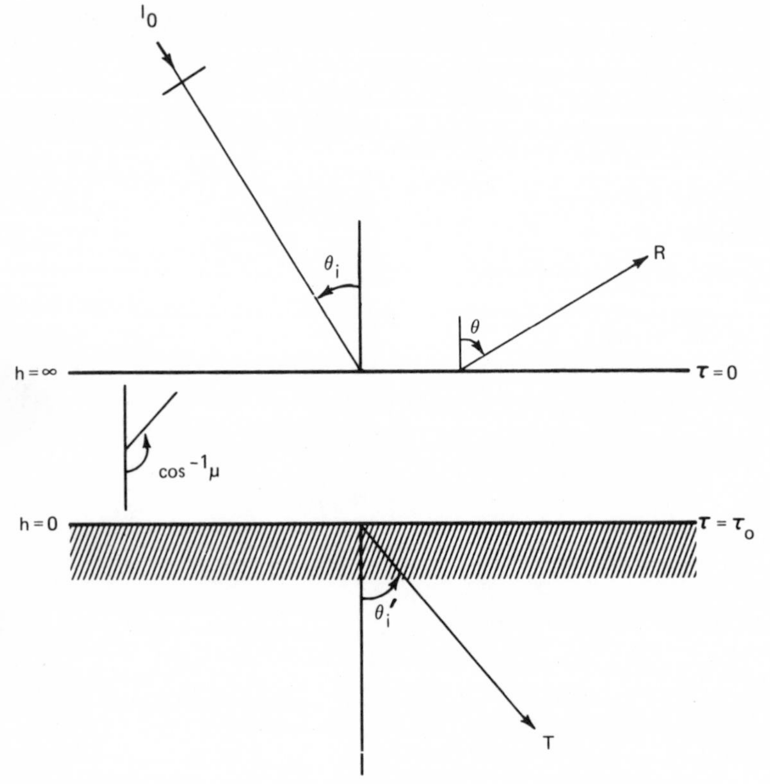

FIG II-12 Definition of zenith angles and optical depth.

The exponential terms are expressions of absorption only, not scattering. The radiation leaving the bottom layer is the sum of the unscattered flux $\mu_s I_0 \exp[-(\zeta_a + \zeta_b)/\mu_i]$ and the diffuse transmissions $\mathbf{T}(\zeta_a + \zeta_b)$. These matrix expressions have functional equivalents for the Stokes parameters (van de Hulst, 1963).

The functions **R**, **T**, **D**, **U**, **S**, and **Q** are 4 row, 4 column matrices, where, for instance,

$$\mathbf{R} = R^{ij}(\mu_s, \mu_i, \varphi - \varphi_i), \qquad i, j = 1, 2, 3, 4 \tag{II-184}$$

and $\mathbf{Z} = \mathbf{X}\,\mathbf{Y}$ is

$$Z^{lj}(\mu_s, \mu_i, \varphi - \varphi_i) = \frac{1}{\pi}\int_0^1 \int_0^{2\pi}\left[\sum_{k=1}^{4} X^{lk}(\mu_s, \mu_i', \varphi - \varphi_i')Y^{kj}(\mu_i', \mu_i, \varphi_i' - \varphi_i)\right]\mu_i'\, d\mu_i'\, d\varphi_i' \tag{II-185}$$

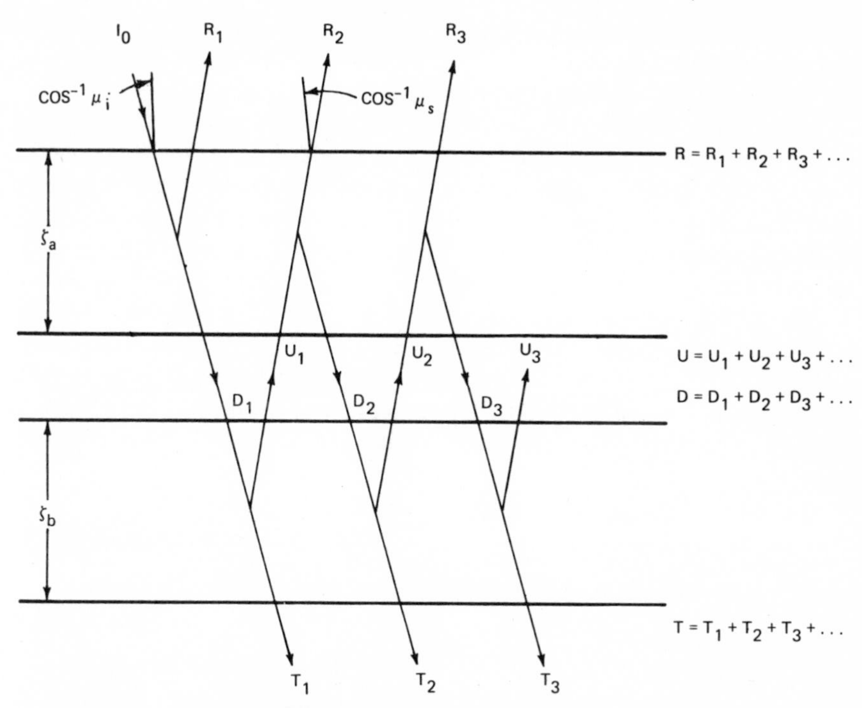

FIG. II-13 Schematic representation of the adding method. The two layers, of optical thicknesses ζ_a and ζ_b, respectively, are physically separated for clarity.

In order to perform the doubling, it is then necessary to treat the original two layers as one equivalent layer with diffuse reflection $\mathbf{R}(\zeta_a + \zeta_b)$ and transmission $\mathbf{T}(\zeta_a + \zeta_b)$ as calculated in Eqs. (II-177) and (II-178). The new optical depth for absorption is the sum of ζ_a and ζ_b.

The calculations are made more expeditiously by expanding the azimuth-dependent terms in a Fourier series expansion of $\varphi - \varphi_\text{i}$. This permits independent treatment of each Fourier series term.

Other techniques for computing multiple scattering are (Hansen and Travis, 1974):

successive orders of scattering,
invariant embedding,
the Monte Carlo method.

The successive orders of scattering method has the property that the intensity is computed for photons scattered once, twice, three times, four times, etc., and the total intensity is the sum over all orders. Integrations may be performed numerically, by using a Gauss–Seidel iterative technique, or by expanding the phase function in Legendre polynomials, and replacing the integral with a finite sum.

The invariant embedding approach is a special case of the doubling technique. The differential equations give the change in reflection and transmission when an optically thin layer is added onto the atmosphere. One variation is known as the method of X and Y functions. This procedure determines certain integral equations for functions dependent upon only one angle and related directly to R and T. These integral equations are then solved numerically.

The Monte Carlo method uses the fact that individual photon scattering is a stochastic process, the phase function being the probability density function for scattering at a particular angle. This approach allows great flexibility when being applied to very complicated problems. The accuracy is limited only by the number of times the game of chance is played by the computer.

NOTES ON SUPPLEMENTARY READING MATERIAL

Born and Wolf (1959): Presents a reasonably complete picture of our present knowledge of optics; discusses the optics of crystals and metals, and diffraction theory.

Chandrasekhar (1960): A presentation of radiative transfer in plane parallel atmospheres as a branch of mathematical physics with its characteristic methods and techniques.

Deirmendjian (1969): Mainly a selected tabulation useful for terrestrial and planetary atmospheric problems, as well as interstellar dust problems; an introduction to single- and multiple-particle scattering.

Mie (1908): Pioneering mathematical description of scattering by spheres.

Newcomb (1960): Although presenting elements of practical and theoretical astronomy, it supplies a reference for useful methods and formulas in spherical geometry.

Ramo and Whinnery (1947): Treats electromagnetic theory in relation to problems in modern radio and electronic engineering; treats many practical applications.

Van de Hulst (1957): A mathematical exposition of scattering with particular emphasis on spherical particles in a wide range of sizes and with varying absorption. Also treats cylinders.

Wendlandt (1968): A collection of papers presented at the American Chemical Society Symposium on Reflectance Spectroscopy, in Chicago, September 1967; papers include instrumentation, theory, and techniques.

Wendlandt and Hecht (1966): A good reference on reflectance spectroscopy, describing instrumentation and applications to a variety of problems.

Wickramasinghe (1967): Presents an account of theories and observations on scattering by the interstellar medium, with particular emphasis on grain theories.

CHAPTER III □ REAL INDEX MEASUREMENT TECHNIQUES FROM ULTRAVIOLET TO INFRARED

A. INTRODUCTION

The refractive (real) portion of the optical complex index of refraction [Eq. (I-1)] is of great value in the identification of liquid or solid materials. The utilization extends from optics, through chemistry, biology, and engineering. Many methods exist for the determination of the refractive portion of the index, and a few of the important techniques will be described. The range of applicability of these techniques is discussed at the end of this chapter.

The required accuracy of determination of the index depends upon the application: minerologists require about ± 0.01 absolute accuracy whereas chemists require an absolute accuracy of ± 0.001, or better. Remote-sensing applications generally require an accuracy between those two limits.

B. REFRACTOMETER

The determination of the real index of refraction of a liquid or solid may be made using a refractometer. Figure III-1 shows a schematic of such a device. The test specimen X is in optical contact with a transparent right prism P. The incident radiation rays A–E are focused on the sample-prism interface. The totally internally reflected rays D' and E' do not enter the viewing telescope T but the refracted rays A', B', C' do. From the geometry, the real index n in terms of the prism index n_p is

$$n = (n_p{}^2 - \sin^2 e)^{1/2} \tag{III-1}$$

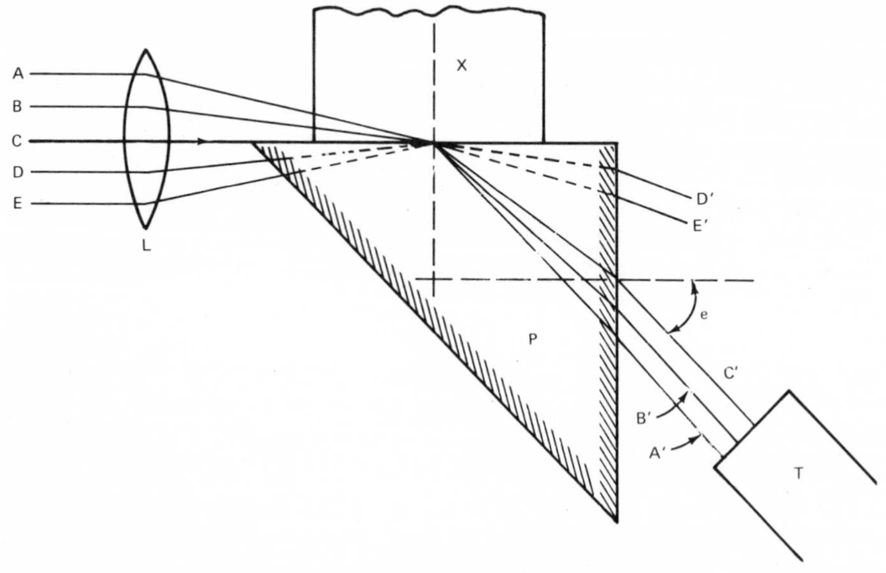

FIG. III-1 Diagram of refractometer.

This technique will not work when $n \geq n_p$. For use in the infrared or ultraviolet, a focusing telescope (and appropriate detector) would be required.

C. MINIMUM ANGLE DEVIATION

Another approach for transparent solid materials is to construct a prism from them, or, for liquids, to use a hollow-cell prism. Essentially, the experimental arrangement is as shown in Fig. III-2. The ray entering the transparent

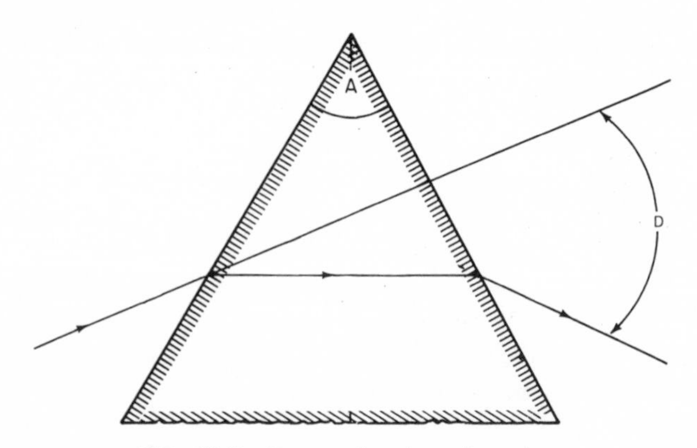

FIG. III-2 Ray passing through a prism.

prism from the left is refracted at entrance and exit. By rotating the prism and the sensor independently, the deviation angle D is minimized to D .

The real index n (for a nonabsorbing material) is given by$_{\mathrm{m}}$.

$$n = \sin \tfrac{1}{2}(D_{\mathrm{m}} + A)/\sin \tfrac{1}{2}A \qquad \text{(III-2)}$$

This technique is useful for any wavelength but requires homogenous specimens large enough for prism construction. The same expression is applicable to a hollow-shell prism, but the walls must exhibit no wedge angle, i.e., must be plane parallel.

D. BECKE LINE

Yet another approach makes use of the fact that transparent solids (with only one index of refraction) disappear from view when immersed in a transparent liquid of exactly the same real index of refraction. From this phenomenon, one may deduce the index of solids (in particulate form) using liquids of known index, or of liquids, using transparent particulates of known index of refraction. This is the Becke line method. A bright line appears on the higher-index side of the boundary between materials of different indices when viewed under a microscope, with appropriate illumination. This bright line moves across the border as the focus of an observing microscope is moved. By observing the behavior of the Becke line on solid particles of unknown index as known index liquids are changed (or known index particles are changed in unknown index liquids), the index may be determined. There are several variations of the optical microscopic technique.

The use of this method becomes difficult for uniaxial or biaxial crystals having two or three indices of refraction. The particles must be properly oriented with respect to each of the two or three axes and the Becke line method applied in each case.

E. FRESNEL'S EQUATIONS

A more general approach involves the use of Fresnel's equations in some form. These equations for reflectance of the parallel ($\hat{\mathbf{R}}_{\parallel}$), and perpendicular ($\hat{\mathbf{R}}_{\perp}$) components are given in Chapter I [Eqs. (I-40) and (I-41)]. By measuring the reflectance of the parallel and/or perpendicular components at several incidence angles, and inverting the equations, the refractive (and absorptive) components may be determined. This procedure is cumbersome, and usually requires a computer. Less accurate, graphical procedures have been described in the literature (Simon, 1951).

F. BREWSTER ANGLE TECHNIQUE

The Brewster angle is defined for transparent dielectric materials as the angle at which an incident beam of unpolarized radiation is reflected as completely polarized. The Brewster angle [Eq. (I-39)] is defined only for nonabsorbing materials. However, for a range of low-absorption-coefficient materials, the angle represented by the minimum reflected parallel component is a pseudo-Brewster angle, and this provides a sufficiently accurate representation of the real index. The determination of the Brewster angle or pseudo-Brewster angle requires a collimated incident beam of radiation and a sensor system that senses collimated reflected (or scattered) radiation. An example of an appropriate system is shown in Fig. III-3 (Egan *et al.*, 1973a).

The system is capable of accurate absolute and relative spectral and polarimetric measurements over the wavelength range from 200 nm to 45 μm, in both the forward scattering and backscattering directions, in the horizontal plane. The source is at the left, the sample at the pivot point *j* and the sensor system at the right.

A high-intensity tungsten iodide lamp *a* is used for measurements between 0.33 and 2.5 μm, a compact xenon arc lamp between 0.2 and 0.35 μm, and a Globar source for the longer wavelength IR. The source radiation is chopped at *b*, passes through condenser *c*, pinhole *d*, and is collimated to within $\frac{1}{2}°$ by lens *e*. All optics are fused quartz for the visible and near UV and KRS-5 for the IR. A polarizer can be located at *g*, and a variable optical attenuator at *f*. A shutter is located at *h*. Measurements of collimated reflected light are made through a range of scattering angles θ. In order to make measurements of the Brewster angle, the sensor system is rotated around the pivot through a series of angle increments and the sample is pivoted to the dotted positions *k′* (from position *k*) for specular reflectance. Then the parallel component specular reflectance (for constant incident radiation) is measured

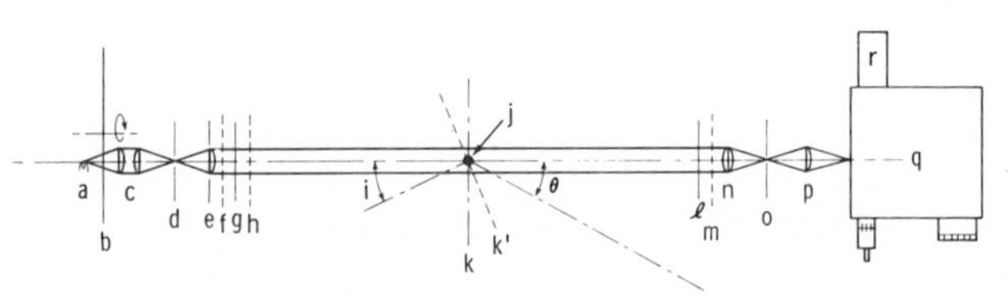

FIG. III-3 Layout of optical system for reflectance measurements. [From Egan *et al.* (1973a).]

for each angular increment of the sensor. The minimum value of the parallel component (zero for a perfect dielectric) occurs at the Brewster angle.

The sensor optics consist of a polarization analyzer *l* and wavelength-selective interference filters as required *m*, objective lenses *n* and field stop *o*, and field lens *p* focusing the radiation on the entrance slit of the monochromator *q*.

The sensor *r* is a photomultiplier between 0.33 and 1.1 μm. For measurements between 1.1 and 4 μm the monochromator *q* is replaced by a filter system and the photomultiplier is replaced by a liquid-nitrogen-cooled indium antimonide detector. From 4 to 12 μm, a liquid-helium-cooled arsenic-doped silicon detector is employed; beyond this zinc- or copper-doped germanium is used to 45 μm.

The sample must be highly polished and flat in order to measure the Brewster angle accurately. If inhomogenous rocks or minerals are measured, there will be an associated surface scattering which prevents the minimum value of the parallel component from reaching zero. The Brewster angle technique is not restricted to any wavelength range, and is a direct determination of the real index of refraction, not dependent on another measurement—the absorption for instance.

A typical Brewster angle measurement result is presented in Fig. III-4. The values for the parallel component are plotted as the ordinate, with the

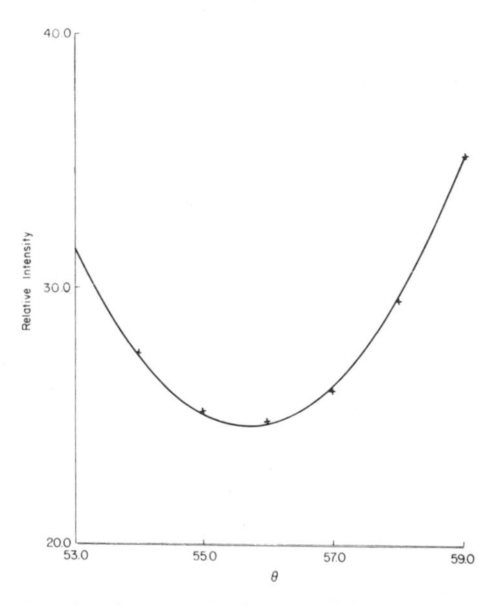

FIG. III-4 Results of a Brewster angle measurement on a compressed limonite pellet at $\lambda = 0.31$ μm.

angle as the abscissa. A least-squares quadratic computer fit and the minimum determined (55.704° as shown, corresponding to $n = 1.466$).

Compressed powder samples are amenable to real index determination by the Brewster angle technique. The primary requirement is that the die used to compress the powder have a highly polished flat face to produce a flat, quasi-specular sample face on the compressed powder sample.

For highly absorbing materials such as semiconductors and metals, the angle of the minimum is the pseudo-Brewster angle. The real and imaginary parts of the optical complex index of refraction must in this case be determined be ellipsometry (Chapter IV).

G. DISCUSSION OF LIMITS OF APPLICABILITY

For homogenous transparent or low-absorption materials, the minimum deviation technique is most accurate (to five places, $\kappa_0 \lesssim 10^{-7}$). When absorption exists, the exact equation [Snell's law, Eq. (I-30)] is indeterminate in the real index unless the absorption coefficient is known. The Becke line and the refractometer techniques are good to four places. The Becke line technique has the additional advantage that it will work with small sample quantities. The Brewster angle technique is good to two or three decimal places but requires that the portion of the beam transmitted into the sample not find its way back out into the sensing optical system. The Fresnel equation technique is analogous to the Brewster angle technique in accuracy.

For inhomogeneous but transparent samples, the results of all techniques are degraded. For a perfect diffuse scatterer, none of the approaches work. Samples exhibiting quasi-specular transmission permit varying degrees of accuracy of real index determination. Of the techniques that have been discussed, the Brewster angle technique is least sensitive to scattering and the Fresnel equation technique (which depends upon absolute reflectivity) is most sensitive to scattering. The other techniques lie between.

For highly absorbing materials, ellipsometry must be used; however, for moderate absorption ($\kappa_0 < 0.01$), either the Brewster angle technique or the Fresnel equation technique may be used. The Becke line technique will work for $\kappa_0 \lesssim 0.002$.

These techniques were originally employed in the visual region. Varying degrees of optical sophistication are required to adapt them to the ultraviolet or infrared. The Becke line method would require an image converter or scanner for microscopic installation. The other approaches would use an appropriate single detector for peaking the signal.

NOTES ON SUPPLEMENTARY READING MATERIAL

Hardy and Perrin (1932): Deals with both pure and applied optics using mathematical and nonmathematical approaches; discusses optical instrumentation, measurement of real portion of index of refraction.

Jenkins and White (1957): Classical physical optics with emphasis on the physical explanation of optical phenomena using graphical means; a good discussion of dielectric and metallic reflection, and polarized light.

CHAPTER IV □ ABSORPTION AND SCATTERING MEASUREMENT TECHNIQUES

A. INTRODUCTION

The variation with wavelength of the optical bulk absorption coefficient α [Eq. (I-5)] of a medium is a unique parameter. As such, it provides the most valuable optical information available for material identification. For transparent or semitransparent nonscattering materials, the specular transmission of a plane parallel slab of a sample may be measured. With a correction for the radiation reflected at the incident and the back surfaces [Eq. (I-3)], the quantity α may be computed; from α, the extinction coefficient κ_0 [Eq. (I-6)] may be computed.

However, most naturally occurring rocks and minerals have scattering, both surface and body, as a result of crystal imperfections. Thus, the transmission of radiation will have some degree of diffuseness, as will the reflection. Because of this, a technique other than specular transmission must be used. A convenient approach to characterize the optical properties of thin sample sections is to measure the total diffuse transmission and total diffuse reflection. Then one may apply a scattering theory such as that of Kubelka and Munk [Eqs. (II-11)–(II-19)] to determine the extinction coefficient κ_0 separate from the scattering s.

B. INTEGRATING SPHERES

The experimental determination of total diffuse transmission and reflection is usually made with an integrating sphere. Currently available commercial integrating spheres (such as the Gier–Dunkle Instruments,

Inc., integrating sphere reflectometer attachment) can make these measurements conveniently between 0.23 and 2.3 μm with a smoked MgO coating.

Another integrating sphere system, using a $BaSO_4$ coating for the region from 0.185 to 2.0 μm and sulfur for the region between 1.5 and 12 μm has been described (Egan and Hilgeman, 1975a). This sphere has the disadvantage of a detector field of view smaller than that for the conventional Gier–Dunkle system, for instance, but it additionally allows the use of cryogenically cooled detectors in the region from 1.1 to 12 μm. The sample measurement geometry is shown in Fig. IV-1.

The measurement procedure for total diffuse transmission (Fig. IV-1a) positions the detector (which has an acceptance angle of 9°) at port *P*2 to view an area of the sphere in front of the diffuser disk, so that its acceptance angle will not be intercepted by the diffuser disk. Port *P*1 is closed with a completely coated plug and *H*2 closed with a mirror. The collimated incident beam enters hole *H*1 and is aimed at the diffuser disk which blocks any direct radiation with no sample in its path from striking the sphere. This forms a reference measurement; the sample then is positioned accurately at entrance hole *H*1, in the incident beam, and the total diffuse absolute transmission is the ratio of this measurement to the first.

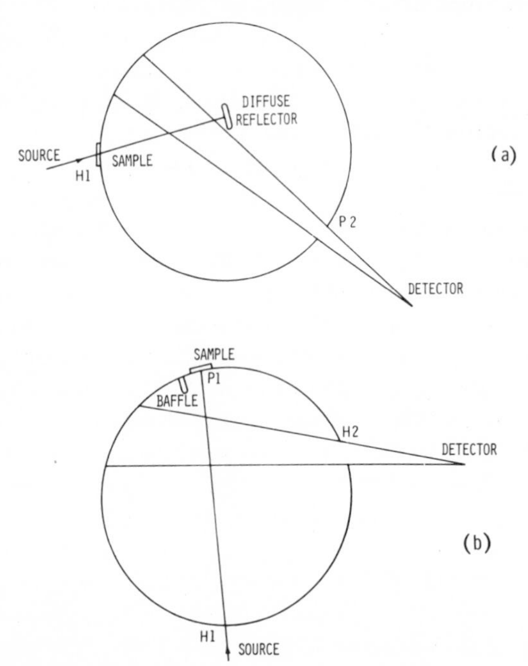

FIG. IV-1 Sample measurement geometry: (a) transmission, (b) reflection. [From Egan and Hilgeman (1975a).]

For total diffuse reflectance (Fig. IV-1b) the measurement procedure involves relocating the detector to hole $H2$ to view the sphere area shielded by the sample baffle, and directing the collimated incident beam through hole $H1$ at the sample. The sample is placed in the sample holder having the vertical baffle which eliminates the first backscatter, making the measurement absolute. This approach is similar to that described by Kortüm (1969).

In theory, only two measurements are necessary to determine total diffuse reflectance. Practically, however, the mismatch of the port and beam sizes requires a correction. Four reflectance measurements are made: (1) with a completely coated 51-mm plug in boss $P1$, (2) with a coated 51-mm plug (that has a 13-mm hole) in boss $P1$, (3) with the 51-mm coated plug (that has the baffle, and hole with the sample) inserted into boss $P1$, and (4) with the 51-mm coated plug (that has the baffle, and hole without the sample) inserted into boss $P1$. The measurements (2) and (4) with the open 13-mm holes serve to correct for the small lack of collimation of the incident radiation. The total diffuse absolute reflectance is given by

$$R = [(3) - (4)]/[(1) - (2)] \qquad \text{(IV-1)}$$

where the parenthesized numbers represent the detector readings under the respective conditions indicated. The sample holder [in reflectance measurement (3) above] is left open behind the sample so that there is no backscatter through the higher-transparency samples. There are also two 13-mm-diameter coated inserts; each may be used as a reference or in the determination of its own total diffuse reflectance.

The instrumentation layout for use with the integrating sphere is shown in Fig. IV-2. The same source optics and detector system may be used as in Fig. III-3.

Certain naturally occurring samples may be quasispecular, and because of the imperfect nature of integrating spheres, errors can occur in the measurement of diffuse transmission (Egan and Hilgeman, 1975a). The error that can occur with an integrating sphere system, the Gier–Dunkle system for instance, in the determination of the true bulk α of a sample has been demonstrated by Egan *et al.* (1973a); in addition, a comparison of the Kubelka–Munk theory [Eqs. (II-11)–(II-19)] was made with a modified Kubelka–Munk (MKM) theory [Eqs. (II-20)–(II-59)]. For this comparison, two well-defined samples were used. Further, an optical system was used that permitted goniometric measurements of total diffuse transmission and reflectance as an absolute reference. This system was the one used for Brewster angle measurements (Fig. III-3), except that the polarizers were omitted.

Verification of the range of application of any flux radiative transfer technique for absorption coefficient determination requires the preparation of well-defined absorption standards. For instance, Corning clear colored

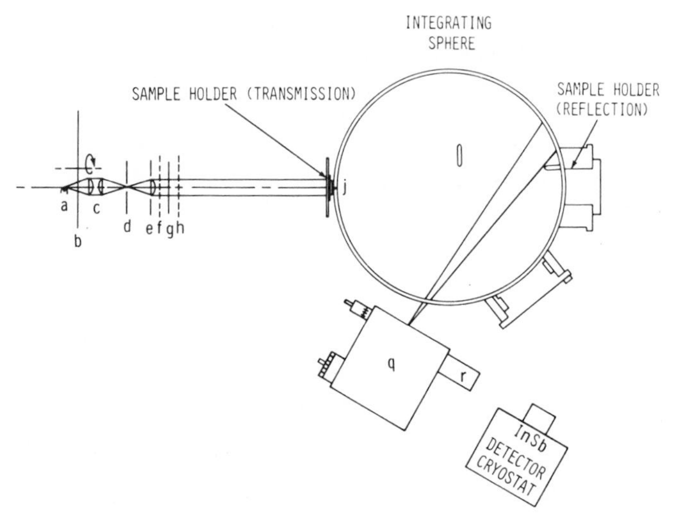

FIG. IV-2 Layout of optical system for integrating sphere. [From Egan and Hilgeman (1975a).]

filter glass might be selected for the presence of a range of specular transmittances between 0.400 and 1.5 μm. Then, having a glass with clearly defined attenuation, it is convenient to subdivide or modify it to introduce scattering centers, without changing the known bulk absorption coefficient.

As a technique to check both theory and apparatus, standard absorbing and scattering samples can be prepared from this glass: for instance, a powder sample, a solid button containing scattering centers such as bubbles, or compressed pellet KBr samples. It is not to be inferred that clear, nonscattering samples should be subdivided purposely in order to obtain the absorption portion of their optical complex index of refraction. Such samples are amenable to standard analyses. The applicability of a radiative transfer approach for absorption coefficient determination is primarily to those materials, such as naturally occurring rocks and minerals, for which bulk samples cannot be obtained without scattering centers. As an example of suitable standards, powder and button samples have been reported on which were prepared from CS1-64 glass (Egan *et al.*, 1973a). This glass is a Corning blue filter with attenuation in the green and red regions and transmission in the near IR. A powder sample was prepared by pulverizing the glass to a size small enough to pass through an ASTM 200 mesh sieve (<74-μm grain sizes). This powder was then retained between two 0.15-mm-thick polished fused quartz microscope slide cover glasses spaced 0.86-mm apart. The boundary conditions for surface reflections appear in the modified Kubelka–Munk theory but not in the Kubelka–Munk theory.

A button sample was prepared by sintering the <74-μm glass powder at a temperature above the softening point ($\sim 725\,°C$) for 1 hour. This produced a sample containing air bubbles with a bubble size distribution given by Fig. IV-3. A large number of fixed spherical scattering centers of small diameter (bubbles) was obtained.

The preparation and analysis of compressed KBr pellet samples will be discussed later in this section.

Once the samples are prepared, the diffuse reflection and transmission must be measured. The most accurate technique is the spectrogoniometric approach using a 4π solid angle for measurement. If the samples have rotational symmetry, measurements in one plane are adequate. The integrating sphere is quite useful for rapid measurements but generally inaccurate for quasi-specular samples. As a first check on the radiative transfer theory, we consider spectrogoniometric measurements on the powder and sintered CS1-64 glass samples. We will subsequently discuss the KBr pellets. Two types of measurements can be made: (1) relative transmission and reflectance, and (2) absolute transmission and reflectance.

(1) For relative reflectance and transmission measurements, a suitable standard must be chosen. One stable surface is USP $MgCO_3$ (prepared by a precipitation process and pressed into blocks). Observations of the diffuse

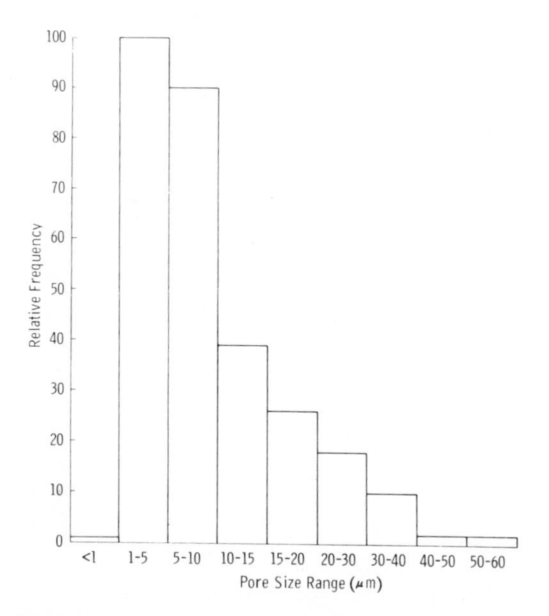

FIG. IV-3 Distribution of pore diameters in a sintered CS1-64 bubble glass. [From Egan *et al.* (1973a).]

reflectance of the $MgCO_3$ and sample for $90° < \theta < 180°$ may be made goniometrically with an instrument such as the one shown in Fig. III-3. These readings, multiplied by $\sin \theta$ and a weighting function decreasing with increasing θ, to account for the actual optical beam size, yield, for instance, the weighted $MgCO_3$ and CS1-64 powder sample reflectance curve (Fig. IV-4). The specular effect of glass cover plates on the powder sample must be added to the diffuse scattering. The ratio of the total area under the weighted diffuse and specular reflectance curves of the CS1-64 powder to the area under the weighted $MgCO_3$ curve yields the relative reflectance of the CS1-64 powder; here, it is 91.3% relative to $MgCO_3$.

The relative transmission is obtained in the same way, as a ratio to the results for $MgCO_3$ of measurements of diffuse transmission of the CS1-64 powder for $0° < \theta < 90°$. The ratio of the weighted diffuse transmission of the CS1-64 powder to the weighted diffuse reflectance of the $MgCO_3$ is the relative transmission; here, it is 0.65%.

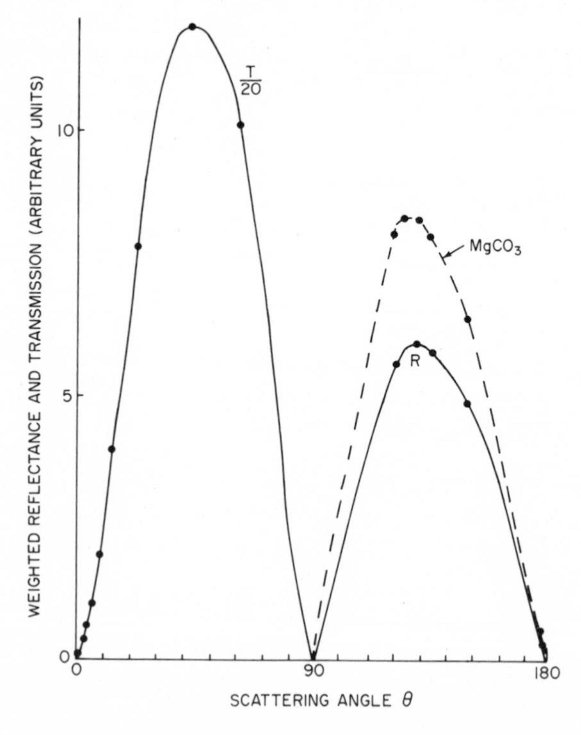

FIG. IV-4 Weighted diffuse scattering curves for $MgCO_3$ and CS1-64 powder standard at 0.400 μm; $T = 0.65\%$ and $R = 91.3\%$ relative to $MgCO_3$ (see text). [From Egan *et al.* (1973a).]

(2) For absolute transmission and reflectance measurements, a reduction to an absolute measurement is made in either of two ways. First, an independent measurement of the absolute diffuse reflectance of $MgCO_3$ in a MgO or $BaSO_4$ coated integrating sphere may be made. (This technique is valid only between 0.25 and 2.3 μm because of absorption out of this region.) Second, the incident beam power may be determined, by transforming it theoretically into a 2π solid scattering angle, as would be done by a perfect diffuser. Then, it may be compared to the measured diffuse scattering observed by a goniometric photometer. (This technique, while less accurate, is usable at all wavelengths.)

Typical results obtained are given in Tables IV-1 and IV-2. Table IV-1 gives a comparison of absolute reflectance and transmission data on the

TABLE IV-1 Corning CS1-64 Glass Absolute Percent of Reflectance (R) and Percent of Transmission (T) Measurements [a]

	Powder				Sintered bubble glass	
	Gier–Dunkle		Modified P-E 13U		Modified P-E 13U	
$\lambda_{\mu m}$	R	T	R	T	R	T
0.400	70.5	0.7	74.8	0.59	44.5	15.2
0.600	57.5	~0.02	55.1	0.02	5.9	2.3
0.907	78.8	1.7	83.2	1.76	26.7	27.5
1.5	70.2	1.0	70.3	0.83	17.8	18.7

[a] From Egan *et al.* (1973a).

TABLE IV-2 Corning CS1-64 Glass Data Analysis [a]

		Powder				Sintered bubble glass			
		KM		Modified KM		KM		Modified KM	
	Bulk								
$\lambda_{\mu m}$	α (cm^{-1})	α[b] (cm^{-1})	s (cm^{-1})	α[b] (cm^{-1})	s (cm^{-1})	α (cm^{-1})	s (cm^{-1})	α (cm^{-1})	s (cm^{-1})
0.400	0.16	7.8	161	2.8	442	1.6	9.5	0.72	23
0.600	9.2	27.1	152	11.6	377	8.3	2.2	13.8	5.0
0.907	0.21	2.8	188	1.15	477	1.7	3.8	1.09	9.1
1.5	0.52	8.2	135	3.2	340	2.9	3.1	2.3	8.3

[a] From Egan *et al.* (1973a).
[b] Corrected for volume fraction.

powder and sintered bubble glass obtained by using a Gier–Dunkle integrating sphere and by the goniometric technique. It can be seen from the comparison that the agreement between techniques is excellent, and is what should be expected. However, it has been noted that for quasi-specular samples, the goniometric technique is generally more accurate (Egan and Hilgeman, 1975a).

Having obtained accurate values for diffuse transmission and reflectance, one may then verify the range of application of radiative transfer models.

For instance, Table IV-2 presents the actual bulk values of the absorption coefficient α as well as the values of α and the scattering s computed with the KM and the modified KM theories. Comparisons are at four wavelengths between blue and the near IR on the CS1-64 powder and sintered glass samples. The actual bulk values of the absorption coefficients α were determined by collimated transmission measurements on a polished clear sample of the glass. For the scattering samples, the values for α and s are based on calculations using the KM theory [Eqs. (II-11)–(II-19)], and the modified KM model [Eqs. (II-20)–(II-59)]; the experimentally determined values for the diffuse R and T for the powder and sintered bubble glass (presented in Table IV-1) form the basis for the calculations. The determinations of Table IV-2 were based on a measured index of refraction of the CS1-64 glass in the wavelength range considered ($n_D = 1.477$). Wavelengths of high transmission of the bulk CS1-64 glass (at 0.400 and 0.907 μm) are compared with wavelengths of low transmission (0.600 μm), and an intermediate transmission wavelength (1.5 μm). It is seen that in all instances the modified Kubelka–Munk scattering model produces values of the absorption coefficient in better agreement with the measured bulk values than those of the KM theory. The scattering coefficients are larger from the modified Kubelka–Munk than from the KM theory. Note that the sintered bubble glass values for α are in closer agreement with the bulk values than for the powder sample, except at 0.600 μm. The divergence between the results for α and the bulk value is attributed to anisotropic scattering such as inequality between forward scattering and backscattering. It is to be expected that the divergence is greater for the powder than the bubble glass because it has much more scattering. In summary, it appears that the absolute values of α calculated using the modified Kubelka–Munk approach are better than those obtained using the KM approach and are higher than the bulk values by amounts ranging from 50% to a factor of 5. While this accuracy is not as good as can be obtained by conventional techniques for clear homogeneous samples, it is sufficient for most purposes.

Frequently, sufficiently thin sections of naturally occurring highly absorbing or friable rocks cannot be made to permit a transmission measurement (e.g., limonite). Then a "powder technique" must be used in which the powder is embedded in a KBr pellet.

In order to apply the powder technique to candidate samples, a powder size comparable to that of a thin light transmitting section must be prepared; a powder sized $\lesssim 1$ μm usually fulfills this requirement. The concentration of this powder should not be too great or there will be no measurable radiation transmitted. On the other hand, if too dilute a sample is used, the sample will be below the multiple-scattering limit, and lose the diffusing property required by the model for analysis.

The 1-μm-sized powders may be prepared in an agate mortar and pestle and the KBr pellets may be prepared, for instance, using a Barnes Engineering Co. Model 126 pellet KBr die.

In order to check the usefulness of this convenient sample preparation technique with the more-accurate modified KM theoretical model, one may select glass standards such as Corning CS1-64, CS5-61, and CS4-77, powder sized <1 μm. Thus, for instance, a quantity of the CS1-64 glass powder added, 10.6% by weight, to the KBr produced a pellet 13 mm in diameter and 1.68-mm thick (Egan *et al.*, 1975). Measurements of diffuse transmission and reflection on this uniformly distributed CS1-64 pellet made using an integrating sphere (Fig. IV-2) and a point-by-point angular technique at 0.400 and 0.600 μm produce very unsatisfactory calculated absorption results. Specular transmission measurements on KBr pellets, which have been used for absorption determinations on atmospheric particulates (Volz, 1972) are even more in error because of the large amount of uncollected scattered radiation. Because of the relative insensitivity of the uniformly distributed sample measurement to absorption, a revised KBr pellet fabrication technique may be applied. The glass and KBr powders can be mixed in a 1:1.6 ratio by weight, for instance, and placed in a layer on one face of the KBr die; the remaining thickness would then be made solely of KBr that acts as a support for the thin mixed layer. The thicknesses resulting in this example would be 0.25 and 1.20 mm for the mixed layer and the KBr layer. Forty percent by weight of the thin layer would be the CS1-64 glass (with an equivalent thickness of 0.11 mm) that had been uniformly distributed in the KBr pellet of the previously described sample. Diffuse transmission and reflection measurements made with the layered pellet improved the calculated absorption results up to a factor of 2 over measurements made with the uniformly distributed sample.

Further checks of the uniform versus layered distributions in the KBr pellets have been made with Corning CS5-61 and CS4-77 glass powders. These glasses have higher bulk absorption at the 0.633-μm wavelength than the CS1-64 glass. In all instances, the layered standards produce the

TABLE IV-3 Comparison of Absorption and Scattering Coefficients for Standard Glass Samples (Contained in KBr Pellets)[a]

				Modified KM theory				
λ	Method[b]	Sample[c]	n_D[d]	α (cm^{-1})	s (cm^{-1})	α_{bulk} (cm^{-1})	$\alpha_{MKM}/\alpha_{bulk}$	α_{bulk}/s ($\times 10^4$)
		CS1-64						
0.400	S	L	1.477	8.00	913	0.16	50.0	1.75
	P	L		3.64	868		22.1	1.84
	S	U		11.76	429		73.5	3.73
	P	U		7.79	301		48.8	5.32
0.600	S	L		12.94	644	9.2	1.41	143
	P	L		8.83	657		0.96	141
	S	U		22.29	287		2.42	320
	P	U		15.71	222		1.71	417
		CS5-61						
0.633	S	L	1.54	21.85	799	14.7	1.5	184
	P	L		24.94	878		1.7	168
	S	U		22.70	191		1.6	770
	P	U		29.43	185		2.0	795
		CS4-77						
	S	L	1.558	59.76	531	37.0	1.6	697
	P	L		55.95	452		1.5	818
	S	U		74.76	106		2.0	3490
	P	U		76.80	70		2.1	5280

[a] From Egan *et al.* (1975).
[b] S, Integrating sphere; P, point-by-point angular measurement.
[c] L, Layered sample; U, uniformly distributed sample.
[d] Sodium D-line index, n_D. (From Corning Color Glass Filter Catalog, Bulletin CFG, 1970.)

better result, with no greater benefit from the more-time-consuming point-by-point measurements than from the integrating sphere measurements. Layered samples have twice the scattering of the uniform samples (Table IV-3) despite the fact that they are made with half as much glass powder. This would suggest, in agreement with Lindberg and Snyder (1973), that the more highly diffusing samples yield more accurate values for the absorption coefficient in a KM-type calculation. This is not to imply that an infinite scattering coefficient produces accurate results; in fact, it can be seen that very high scattering to absorption ratios produce inaccuracies (Table IV-3).

However, poorest agreement with the bulk absorption coefficient occurs, even with the layered high-scattering sample, when the absorption is low (CS1-64 glass at 0.400 μm). The accuracy of the determination of the absorption coefficient α with absorption coefficient and with the ratio of absorption to scattering coefficients is also shown in Table IV-3. For a given preparation technique (layered L or uniform U) and measurement method (integrating sphere S or goniometric P) as the bulk absorption increases, and as the ratio of bulk absorption to scattering increases, the value of the theoretically derived absorption coefficient from the modified KM model approaches 1.5 times the bulk value. This comparison can be made because the amount of glass included in each type of sample (uniform or layered) is nearly equal. This limiting value of 1.5 is approached for bulk $\alpha > 10\ \mathrm{cm}^{-1}$ and bulk $\alpha/s > 300$. The upper bound on α determined by this technique would be expected to occur when the sample particle size was larger than the optical depth (skin depth).

The size distribution of the sample particles is not critical to the results, because the effect of particle size will primarily be to change the scattering. Further, it is expected that the difference between the index of refraction of KBr [varying from 1.559 in the visible to 2.148 at 0.2 μm; Weast and Selby (1966), Wolfe (1965)] and that of the sample will contribute to additional scattering effects rather than absorption.

From the preceding discussion, it can be seen that measurement of the absorption coefficient of inhomogeneous materials is not as accurate as for homogeneous materials, but, within certain limited ranges of α and s, reasonable values can be obtained. The integrating sphere measurements are more convenient, but the goniometric technique is more general and accurate.

C. ELLIPSOMETRY

"Ellipsometry" describes reflection polarimetry or reflection spectroscopy, and is the measurement of the state of polarization of radiation reflected from a surface. The technique is applied to the determination of the optical complex index of refraction of strongly absorbing nonscattering solids and surface films. The mathematical basis for ellipsometry lies in the determination of the phase shift Δ [Eq. (I-42)] between the parallel and perpendicular components of radiation reflected from a surface. An example of the apparatus for the measurement is that shown in Fig. III-3 with the addition of a Babinet–Soleil compensator just in front of the polarization analyzer l. The measurement requires the determination of the principal azimuthal angle $\hat{\psi}_i$ [Eq. (I-43)] and the phase shift Δ. With incident radiation plane polarized at 45° to the plane of incidence, equal and phase coherent components exist for the parallel and perpendicular waves. Upon reflection of the incident radiation, a phase shift Δ occurs between the two components that is dependent upon the angle of incidence θ_i. In analogy with the Brewster angle, the principal angle of incidence $\hat{\theta}_i$ occurs where the polarization of the parallel reflected component is minimum. The principal azimuth $\hat{\psi}_r$ [from Eq. (I-42)] at the angle $\hat{\theta}_i$ is given by the ratio of the parallel and perpendicular reflected components. The principal azimuthal angle $\hat{\psi}_i$ may also be determined with a Babinet–Soleil compensator, because the phase shift Δ is $\pi/2$ at the principal angle. Thus, by setting the Babinet–Soleil compensator to retard the parallel wave relative to the perpendicular wave by $\pi/2$, the waves are placed in time phase, and the principal azimuth is given by the angular position of the analyzer relative to plane of incidence.

The absorptive and refractive components are then given approximately as

$$\kappa_0/n = \tan 2\hat{\psi}_r \tag{IV-2}$$

$$n = \sin \hat{\theta}_i \tan \hat{\theta}_i/[1 + (\kappa_0/n)^2]^{1/2} \tag{IV-3}$$

The exact expressions are given by Geiger and Scheel (1928).

NOTES ON SUPPLEMENTARY READING MATERIAL

Born and Wolf (1959): Discusses optics of metals and crystals, thin films, diffraction by a sphere in terms of the Mie formulas, scattering, and extinction.

Jenkins and White (1957): Optics of dielectrics, metals, and crystals with emphasis on the physical explanation of optical phenomena using graphical means.

Hardy and Perrin (1932): Deals with pure and applied optics using mathematical and nonmathematical approaches; discusses instrumentation for making reflectance measurements.

Stratton (1941): Treats variable electromagnetic fields and theory of wave propagation from fundamental concepts to application; discusses elliptical polarization and scattering of electromagnetic radiation by a sphere.

Wendlandt (1968): A collection of papers from the American Chemical Society Symposium on Reflectance Spectroscopy, September 1967, in Chicago; a number of papers on diffuse reflectance measurements and on instrumentation.

Wendlandt and Hecht (1966): The two-flux scattering model is presented along with formulas relating scattering and absorption to reflectance and transmittance.

CHAPTER V □ OPTICAL COMPLEX INDEX OF REFRACTION BETWEEN 0.185 AND 2.6 μm

A. INTRODUCTION

The key to modeling naturally occurring rock and mineral surfaces lies in the optical complex index of refraction. Very little information has been available in the technical and scientific literature about appropriate values, particularly over a wide spectral range, from the ultraviolet into the infrared. A selection of optical complex indices of refraction from 0.185 to 2.6 μm would be useful for modeling terrestrial and planetary surfaces. With this objective in mind, we have assembled optical complex indices of refraction of 24 naturally occurring rocks and minerals that are of general interest. The indices have been obtained using measuring techniques described in Chapters III and IV. About half of the data has not previously been published.

The measurements involve the use of the Brewster angle technique (Chapter III) for the real (refractive) portion of the index of refraction and observations of total diffuse transmission and reflection with the modified Kubelka–Munk (MKM) scattering theory to determine the imaginary (absorption) portion κ_0. The samples were either thin sections (TS) or compressed pellets (CP).

Table V-1 is a list of the samples together with the chemical formula representing the composition, the source, and the table or figures in this text in which the complex index data are located.

Chemical analyses exist on some of the samples. For instance, there are two Table Mountain basalt samples that differ in the amount of iron, revealing variations even within the same sample collection. There are also differences in the optical complex indices of refraction. Table V-2 lists the chemical analyses of several of these samples.

Tables V-1 and V-2, and Tables V-3–V-26 with corresponding Figs. V-1–V-24 follow the Notes on Supplementary Reading Material.

B. EXAMPLES

Column captions (Tables V-3–V-26 and Figs. V-1–V-24) designate the wavelength λ (micrometers), the log base 10 of the total diffuse transmission (T) the total diffuse reflection (R), the log base 10 of the imaginary portion of the index of refraction (computed from T and R) and the real portion of the index of refraction n. The complex indices of troilite and pyrrhotite were determined on bulk (thick) samples by ellipsometry.

The thicknesses of the thin sections are indicated above the tabular data; the compressed pellets were prepared as described in Chapter IV from mixtures of the respective specimens with KBr powder in approximately equal quantities by weight (except for highly absorbing samples when less sample was used) to form thin layers with effective sample thicknesses as shown.

The results are plotted on pages facing the tabular data for each sample. The samples include pyroxenes, feldspars, clays, basalts, conglomerates, and iron compounds. The selection was based on various remote sensing applications ranging from terrestrial, lunar, and Martian surfaces to interstellar space.

NOTES ON SUPPLEMENTARY READING MATERIAL

Burns (1970): Nonmathematical exposition to develop an understanding of the role played by transition metals in silicate minerals and also in geological processes.

Hodgman (1950): Presents, in a concise form, a tremendous amount of accurate and reliable data in the fields of chemistry and physics; more specifically, on the optical properties of materials.

Hurlbut (1959): The crystal–chemical approach is used to relate the optical properties of minerals through fundamental structural and chemical considerations; many minerals are discussed.

Kerr (1959): Text intended primarily for use in thin-section studies, but descriptions and tables are presented applicable to the study of mineral fragments; real portion of optical complex indices of refraction for many minerals included.

Seitz (1940): Presents theoretical and experimental aspects of the electronic structure of solid bodies that can affect the optical complex index of refraction.

TABLE V-1 List of Samples

Type of material	Sample	Chemical formula or composition	Source	Table No.	Figure No.
Pyroxenes	Augite	$(Ca, Na)(Mg, Fe'', Fe''', Al)(Si, Al)_2O_6$	Canada	V-5	V-3
	Diopside	$CaMg(Si_2O_6)$	India	V-11	V-9
	Enstatite	$Mg_2(Si_2O_6)$	India	V-12	V-10
Feldspar	Bytownite	$Ab_{30}An_{70}$ to $(Ab_{10}An_{90})$; $Ab \equiv Na(AlSi_3O_8)$; $An \equiv Ca(Al_2Si_2O_8)$	Minnesota	V-10	V-8
Iron Compounds	Limonite	$FeO(OH) \cdot nH_2O$	Venango County, Pennsylvania	V-16	V-14
	Pyrrhotite	$Fe_{0.93}S$	Zacatecas, Mexico	V-25	V-23
	Troilite	FeS	Del Norte, California	V-24	V-22
Clay	Montmorillonite	$(Al, Mg)_8(Si_4O_{10})_3 \cdot (OH)_{10} \cdot 12H_2O$	Amory, Mississippi	V-20	V-18
	Montmorillonite	$(Al, Mg)_8(Si_4O_{10})_3 \cdot (OH)_{10} \cdot 12H_2O$	Clay Spur, Wyoming	V-19	V-17
	Kaolinite	$Al_4(Si_4O_{10})(OH)_8$	Macon, Georgia	V-15	V-13
	Illite	$KAl_2(OH)_2[AlSi_3(O, OH)_{10}]$	Fithian, Illinois	V-14	V-12

Rocks and conglomerates	Basalt	Pyroxene + plagioclase feldspar	Chimney Rock, Watchung, New Jersey	V-6	V-4
	Basalt	Pyroxene + plagioclase feldspar	Table Mountain, Golden, Colorado	V-8	V-6
	Basalt	Pyroxene + plagioclase feldspar	Table Mountain, Golden, Colorado	V-9	V-7
	Basalt (Columbia River)	Pyroxene + plagioclase feldspar	Bridal Veil Quad, Oregon	V-7	V-5
	Andesite	Ferromagnesium minerals	Guano Valley, Lake County, Oregon	V-3	V-1
	Andesite	Ferromagnesium minerals	Volcano Tunaba, Limon, Costa Rica	V-4	V-2
	Shale	Aluminium silicate	Pierre shale	V-23	V-21
	Bruderheim meteorite	Olivine, hypersthene chondritic (Pellet)	Canada	V-17	V-15
	Bruderheim meteorite	Olivine, hypersthene chondritic (thin section)	Canada	V-18	V-16
	Granodiorite	Plagioclase feldspar + ferromagnesium minerals	Westerly Granite, Rhode Island	V-13	V-11
	Rhyolite welded tuff	Similar in composition to granite	Mt. Rogers, Virginia	V-26	V-24
	Obsidian	Chemically the same as rhyolite	Glass Mt., Siskiyou County, California	V-21	V-19
	Mica schist	Mica + quartz	France	V-22	V-20

TABLE V-2 Chemical Analyses of Selected Rock and Mineral Samples[a]

Chemical constituents[b]	Table mountain basalt (Sample 1)	Table mountain basalt (Sample 2)	Columbia river basalt	Chimney rock basalt (N.J.)	Andesite (Guano valley, Or.)	Andesite (Volcano Tunaba, Costa Rica)	Montmorillonite (Amory, Mississippi)
SiO_2	53.30	53.05	53.55	51.45	58.60	55.10	50.15
Al_2O_3	16.12	16.12	13.78	13.94	16.43	16.42	16.39
Fe_2O_3	5.06	5.35	2.75	2.77	4.89	5.93	7.27
FeO	3.91	3.63	9.72	7.48	1.75	1.92	2.59
MgO	3.65	3.71	3.51	7.56	3.02	5.41	2.51
CaO	6.42	6.49	6.82	11.02	4.60	7.45	2.40
Na_2O	3.39	3.44	3.39	2.24	4.46	3.66	0.13
K_2O	4.29	4.31	1.66	0.26	2.71	2.14	1.13
H_2O^+	2.26	2.27	1.57	0.95	1.72	0.24	6.85
H_2O^-				0.46		0.02	6.14
CO_2	0.15	0.13	0.02	0.16	0.04	0.11	2.21
TiO_2	0.87	0.88	2.38	1.04	1.09	0.96	0.68
P_2O_5	0.56	0.58	0.41	0.19	0.52	0.42	1.04
MnO	0.19	0.20	0.21	0.18	0.11	0.14	0.31
S		0.010	0.037		0.00		
Fe							
FeS							
Cr							
Ni							
Zn							
Co							
Cu							
Cr_2O_3							
Pb							
C							
Other volatiles	—	—	—	—	—	—	—
Total	100.17	100.17	99.79	99.70	99.94	99.72	99.80

[a] Data taken, in part, from Egan and Hilgeman (1975b) and Egan *et al.* (1975).
[b] Concentrations of chemical constituents are given in percent.
[c] Loss on ignition, 14.38%.

Limonite (Venango County, Pa.)	Bruderheim meteorite	Bytownite (Minnesota)	Augite (Canada)	Enstatite (India)	Diopside (India)	Troilite (Del Norte County, California)	Pyrrhotite (Zacatecas, Mexico)
5.10	39.94	47.75	52.50	53.60	50.60	1.40	3.45
4.94	1.86	32.30	0.28	0.81	0.80	0.62	1.46
73.14	—	0.06	2.64	3.88	3.74	—	—
0.16	12.94	0.28	4.20	9.76	2.56	—	—
0.36	24.95	0.43	14.28	30.50	16.13	2.07	0.36
0.15	1.74	15.69	25.00	0.44	24.95	0.25	3.76
0.05	1.01	2.54	0.36	0.05	0.32	0.02	0.03
0.52	0.13	0.08	0.01	0.01	0.005	0.00	0.01
10.75[c]	0.10	0.54	0.24	0.16	0.22	—	—
1.74[c]	0.01					—	—
0.55[c]	—	0.15	0.10	0.18	0.08	—	1.65
0.13	0.12	0.03	0.01	0.01	0.13	0.00	0.00
0.54	0.29	0.01	0.02	0.05	0.02	—	—
0.25	0.33	0.03	0.21	0.21	0.16	0.045	1.09
	—					34.01	32.04
	8.59					58.88	52.22
	6.38					—	—
	—					0.25	—
	1.30					0.28	0.000
	—					0.05	0.015
	0.05					—	—
	—					0.25	0.016
	0.60					0.091	0.001
	—					0.25	0.09
	0.04					—	—
1.34	—	—	—	—	—	—	—
99.72	100.38	99.89	99.85	99.66	99.71	98.22	96.19

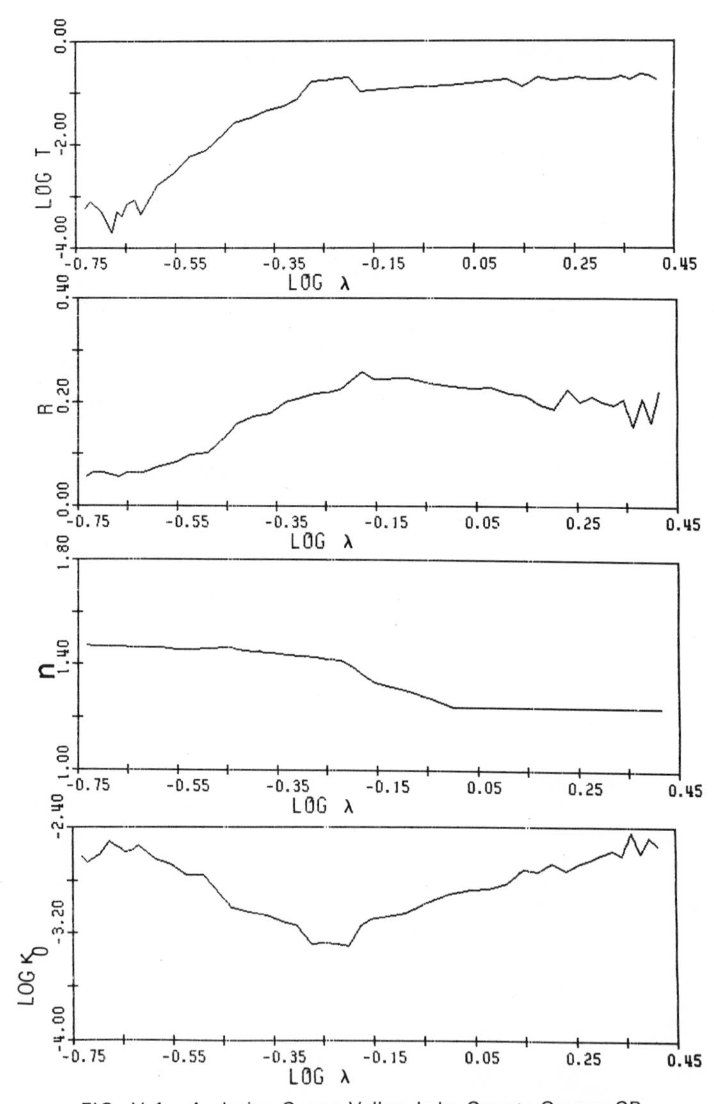

FIG. V-1 Andesite, Guano Valley, Lake County Oregon CP

TABLE V-3 Andesite, Guano Valley, Lake County, Oregon CP[a,b]

LAMBDA	LOG T	R	LOG K_o	n
0.185	-.323E+01	0.056	-.262E+01	1.470
0.190	-.311E+01	0.064	-.266E+01	1.469
0.200	-.331E+01	0.063	-.261E+01	1.469
0.210	-.372E+01	0.057	-.251E+01	1.469
0.215	-.329E+01	0.055	-.254E+01	1.468
0.220	-.339E+01	0.062	-.255E+01	1.467
0.225	-.314E+01	0.064	-.259E+01	1.466
0.233	-.307E+01	0.063	-.258E+01	1.466
0.240	-.336E+01	0.065	-.253E+01	1.465
0.260	-.279E+01	0.077	-.264E+01	1.465
0.280	-.255E+01	0.084	-.268E+01	1.457
0.300	-.224E+01	0.099	-.276E+01	1.457
0.325	-.211E+01	0.101	-.276E+01	1.461
0.360	-.170E+01	0.144	-.295E+01	1.465
0.370	-.158E+01	0.157	-.301E+01	1.457
0.400	-.148E+01	0.172	-.304E+01	1.448
0.433	-.134E+01	0.179	-.307E+01	1.443
0.466	-.125E+01	0.200	-.311E+01	1.438
0.500	-.111E+01	0.208	-.315E+01	1.433
0.533	-.783E+00	0.215	-.328E+01	1.427
0.566	-.754E+00	0.218	-.327E+01	1.421
0.600	-.712E+00	0.224	-.328E+01	1.415
0.633	-.701E+00	0.241	-.329E+01	1.390
0.666	-.971E+00	0.258	-.315E+01	1.360
0.700	-.951E+00	0.243	-.309E+01	1.330
0.817	-.893E+00	0.246	-.304E+01	1.300
0.907	-.879E+00	0.235	-.297E+01	1.270
1.000	-.860E+00	0.230	-.291E+01	1.240
1.105	-.830E+00	0.226	-.287E+01	1.240
1.200	-.772E+00	0.227	-.287E+01	1.240
1.303	-.740E+00	0.216	-.283E+01	1.240
1.400	-.873E+00	0.211	-.272E+01	1.240
1.500	-.710E+00	0.195	-.274E+01	1.240
1.600	-.754E+00	0.187	-.267E+01	1.240
1.700	-.730E+00	0.223	-.273E+01	1.240
1.800	-.693E+00	0.200	-.268E+01	1.240
1.900	-.740E+00	0.209	-.265E+01	1.240
2.000	-.744E+00	0.200	-.261E+01	1.240
2.100	-.740E+00	0.194	-.257E+01	1.240
2.200	-.678E+00	0.205	-.261E+01	1.240
2.300	-.747E+00	0.153	-.244E+01	1.240
2.400	-.646E+00	0.206	-.260E+01	1.240
2.500	-.660E+00	0.162	-.248E+01	1.240
2.600	-.735E+00	0.221	-.254E+01	1.240

[a] Effective thickness, 0.00342 cm.

[b] Data taken, in part, from Egan *et al.* (1975).

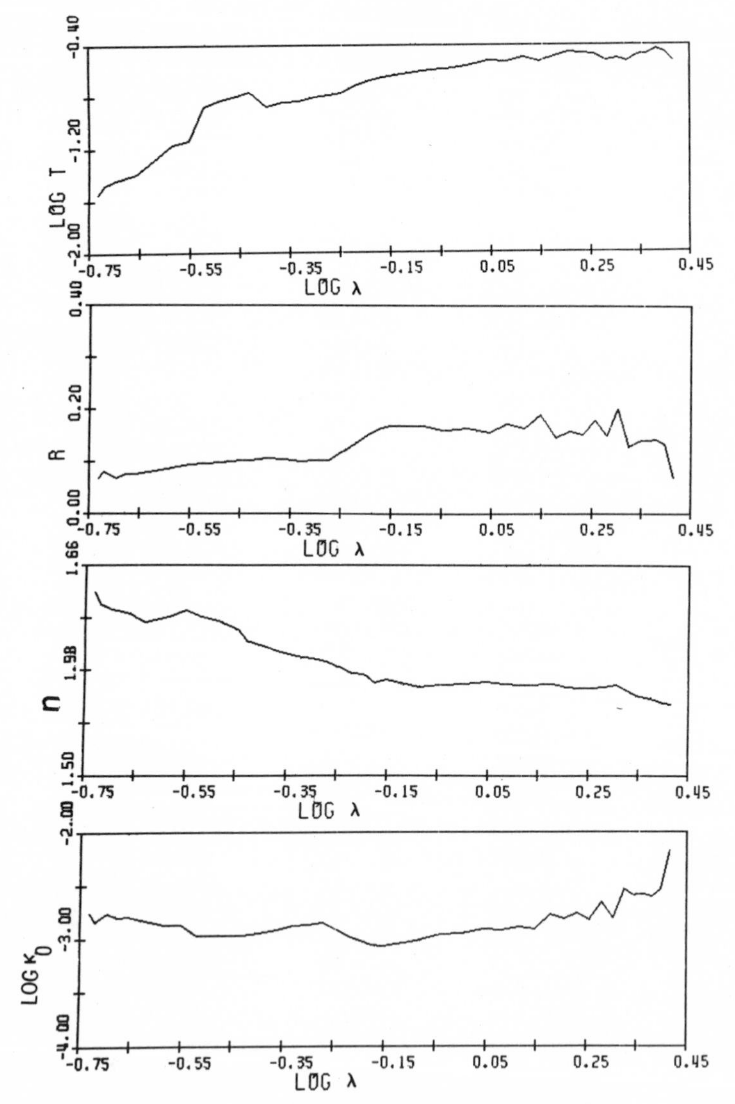

FIG. V-2 Andesite, Volcano Tunaba, Costa Rica CP.

TABLE V-4 Andesite, Volcano Tunaba, Costa Rica CP[a]

LAMBDA	LOG T	R	LOG K_0	n
0.185	-.155E+01	0.068	-.276E+01	1.639
0.190	-.149E+01	0.079	-.285E+01	1.630
0.200	-.144E+01	0.067	-.276E+01	1.626
0.210	-.142E+01	0.075	-.280E+01	1.624
0.220	-.139E+01	0.075	-.280E+01	1.622
0.233	-.132E+01	0.079	-.282E+01	1.616
0.260	-.118E+01	0.087	-.287E+01	1.620
0.280	-.114E+01	0.092	-.288E+01	1.625
0.300	-.879E+00	0.095	-.297E+01	1.620
0.325	-.827E+00	0.097	-.298E+01	1.617
0.355	-.788E+00	0.101	-.298E+01	1.611
0.370	-.764E+00	0.101	-.298E+01	1.602
0.400	-.873E+00	0.105	-.295E+01	1.598
0.433	-.842E+00	0.103	-.293E+01	1.593
0.466	-.833E+00	0.100	-.289E+01	1.590
0.500	-.807E+00	0.101	-.287E+01	1.588
0.533	-.785E+00	0.101	-.286E+01	1.586
0.566	-.775E+00	0.118	-.291E+01	1.582
0.600	-.714E+00	0.134	-.298E+01	1.577
0.633	-.686E+00	0.149	-.303E+01	1.576
0.666	-.660E+00	0.161	-.306E+01	1.570
0.700	-.640E+00	0.168	-.307E+01	1.572
0.817	-.604E+00	0.167	-.303E+01	1.567
0.907	-.590E+00	0.160	-.297E+01	1.568
1.000	-.567E+00	0.164	-.296E+01	1.569
1.105	-.524E+00	0.156	-.292E+01	1.571
1.200	-.529E+00	0.171	-.293E+01	1.569
1.303	-.499E+00	0.164	-.289E+01	1.568
1.400	-.530E+00	0.189	-.291E+01	1.568
1.500	-.493E+00	0.145	-.277E+01	1.569
1.600	-.462E+00	0.158	-.282E+01	1.567
1.700	-.466E+00	0.152	-.277E+01	1.566
1.800	-.472E+00	0.179	-.282E+01	1.566
1.900	-.524E+00	0.148	-.266E+01	1.567
2.000	-.510E+00	0.201	-.281E+01	1.568
2.100	-.529E+00	0.128	-.254E+01	1.564
2.200	-.483E+00	0.138	-.259E+01	1.560
2.300	-.472E+00	0.140	-.259E+01	1.559
2.400	-.435E+00	0.142	-.261E+01	1.557
2.500	-.457E+00	0.132	-.254E+01	1.555
2.600	-.532E+00	0.068	-.218E+01	1.554

[a] Effective thickness, 0.0025 cm.

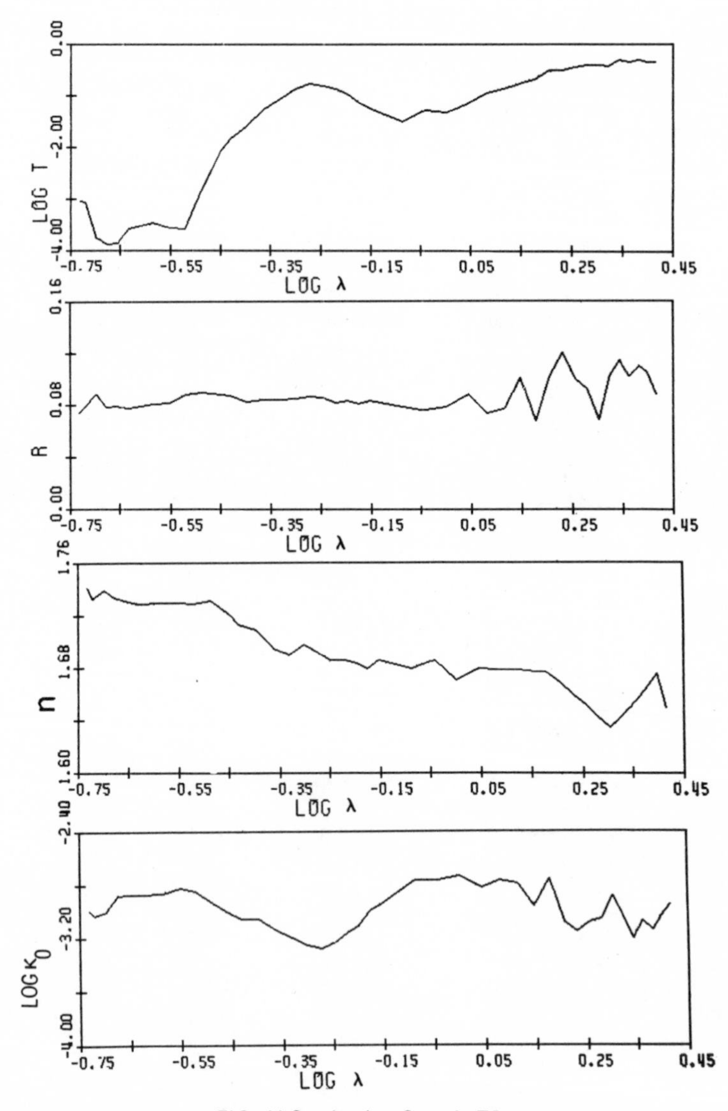

FIG. V-3 Augite, Canada TS.

TABLE V-5 Augite, Canada TS[a,b]

LAMBDA	LOG T	R	LOG K_0	n
0.185	-.304E+01	0.073	-.300E+01	1.741
0.190	-.307E+01	0.078	-.304E+01	1.732
0.200	-.377E+01	0.088	-.300E+01	1.739
0.210	-.389E+01	0.078	-.289E+01	1.733
0.220	-.388E+01	0.079	-.288E+01	1.731
0.233	-.358E+01	0.077	-.288E+01	1.728
0.260	-.348E+01	0.081	-.287E+01	1.729
0.280	-.357E+01	0.081	-.283E+01	1.729
0.300	-.359E+01	0.087	-.285E+01	1.728
0.325	-.282E+01	0.090	-.294E+01	1.731
0.355	-.207E+01	0.088	-.303E+01	1.720
0.370	-.184E+01	0.087	-.306E+01	1.712
0.400	-.160E+01	0.082	-.306E+01	1.708
0.433	-.126E+01	0.083	-.315E+01	1.694
0.466	-.106E+01	0.084	-.320E+01	1.689
0.500	-.860E+00	0.085	-.326E+01	1.697
0.533	-.780E+00	0.086	-.329E+01	1.691
0.566	-.801E+00	0.085	-.325E+01	1.685
0.600	-.863E+00	0.081	-.316E+01	1.685
0.633	-.963E+00	0.082	-.311E+01	1.683
0.666	-.113E+01	0.080	-.300E+01	1.679
0.700	-.125E+01	0.083	-.295E+01	1.685
0.817	-.151E+01	0.078	-.277E+01	1.679
0.907	-.130E+01	0.076	-.277E+01	1.685
1.000	-.134E+01	0.078	-.274E+01	1.670
1.105	-.114E+01	0.087	-.282E+01	1.679
1.200	-.947E+00	0.073	-.277E+01	1.678
1.303	-.873E+00	0.077	-.280E+01	1.678
1.400	-.775E+00	0.101	-.297E+01	1.677
1.500	-.686E+00	0.067	-.276E+01	1.676
1.600	-.509E+00	0.101	-.308E+01	1.668
1.700	-.500E+00	0.120	-.316E+01	1.659
1.800	-.444E+00	0.108	-.308E+01	1.651
1.900	-.405E+00	0.092	-.306E+01	1.642
2.000	-.407E+00	0.068	-.289E+01	1.634
2.100	-.423E+00	0.102	-.305E+01	1.641
2.200	-.314E+00	0.114	-.320E+01	1.649
2.300	-.348E+00	0.102	-.308E+01	1.657
2.400	-.309E+00	0.110	-.314E+01	1.666
2.500	-.353E+00	0.105	-.304E+01	1.675
2.600	-.343E+00	0.088	-.296E+01	1.649

[a] Effective thickness, 0.01 cm.

[b] Data taken, in part, from Egan and Hilgeman (1975b).

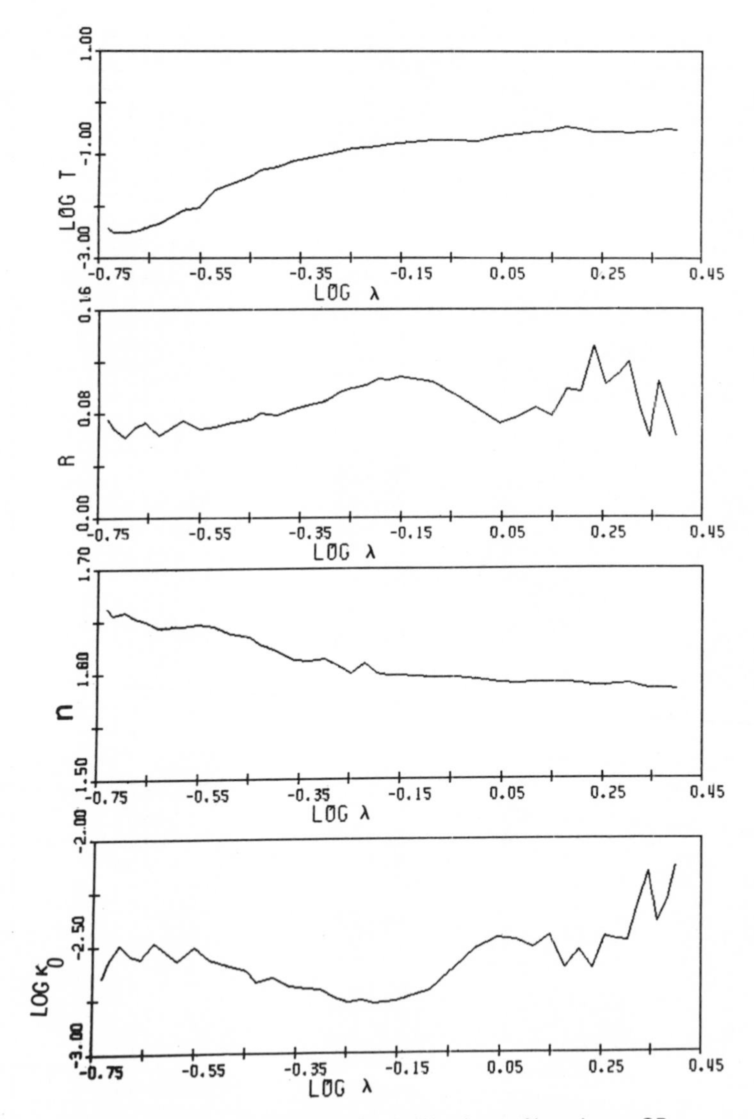

FIG. V-4 Basalt, Chimney Rock, Watchung, New Jersey CP.

TABLE V-6 Basalt, Chimney Rock, Watchung, New Jersey CP[a]

LAMBDA	LOG T	R	LOG K_o	n
0.185	-.243E+01	0.076	-.265E+01	1.662
0.190	-.250E+01	0.068	-.258E+01	1.655
0.200	-.250E+01	0.062	-.250E+01	1.658
0.210	-.250E+01	0.070	-.255E+01	1.652
0.220	-.244E+01	0.072	-.256E+01	1.649
0.233	-.235E+01	0.063	-.249E+01	1.643
0.260	-.209E+01	0.074	-.257E+01	1.644
0.280	-.202E+01	0.068	-.251E+01	1.646
0.300	-.169E+01	0.069	-.257E+01	1.644
0.325	-.158E+01	0.073	-.259E+01	1.637
0.355	-.142E+01	0.075	-.262E+01	1.634
0.370	-.131E+01	0.079	-.267E+01	1.627
0.400	-.125E+01	0.078	-.265E+01	1.621
0.433	-.113E+01	0.083	-.269E+01	1.612
0.466	-.107E+01	0.086	-.270E+01	1.611
0.500	-.101E+01	0.088	-.271E+01	1.613
0.533	-.951E+00	0.096	-.275E+01	1.606
0.566	-.893E+00	0.099	-.277E+01	1.599
0.600	-.870E+00	0.101	-.276E+01	1.609
0.633	-.851E+00	0.106	-.277E+01	1.599
0.666	-.810E+00	0.105	-.277E+01	1.597
0.700	-.799E+00	0.108	-.276E+01	1.597
0.817	-.733E+00	0.104	-.271E+01	1.595
0.907	-.735E+00	0.093	-.261E+01	1.595
1.000	-.745E+00	0.083	-.252E+01	1.593
1.105	-.646E+00	0.071	-.247E+01	1.590
1.200	-.620E+00	0.077	-.248E+01	1.589
1.303	-.575E+00	0.084	-.251E+01	1.590
1.400	-.559E+00	0.077	-.246E+01	1.590
1.500	-.461E+00	0.098	-.261E+01	1.590
1.600	-.520E+00	0.097	-.253E+01	1.589
1.700	-.564E+00	0.131	-.261E+01	1.587
1.800	-.572E+00	0.102	-.246E+01	1.587
1.900	-.561E+00	0.109	-.248E+01	1.588
2.000	-.583E+00	0.119	-.248E+01	1.589
2.100	-.569E+00	0.083	-.231E+01	1.586
2.200	-.567E+00	0.061	-.216E+01	1.584
2.300	-.532E+00	0.104	-.240E+01	1.584
2.400	-.509E+00	0.082	-.229E+01	1.584
2.500	-.535E+00	0.062	-.214E+01	1.583

[a] Effective thickness, 0.0028 cm.

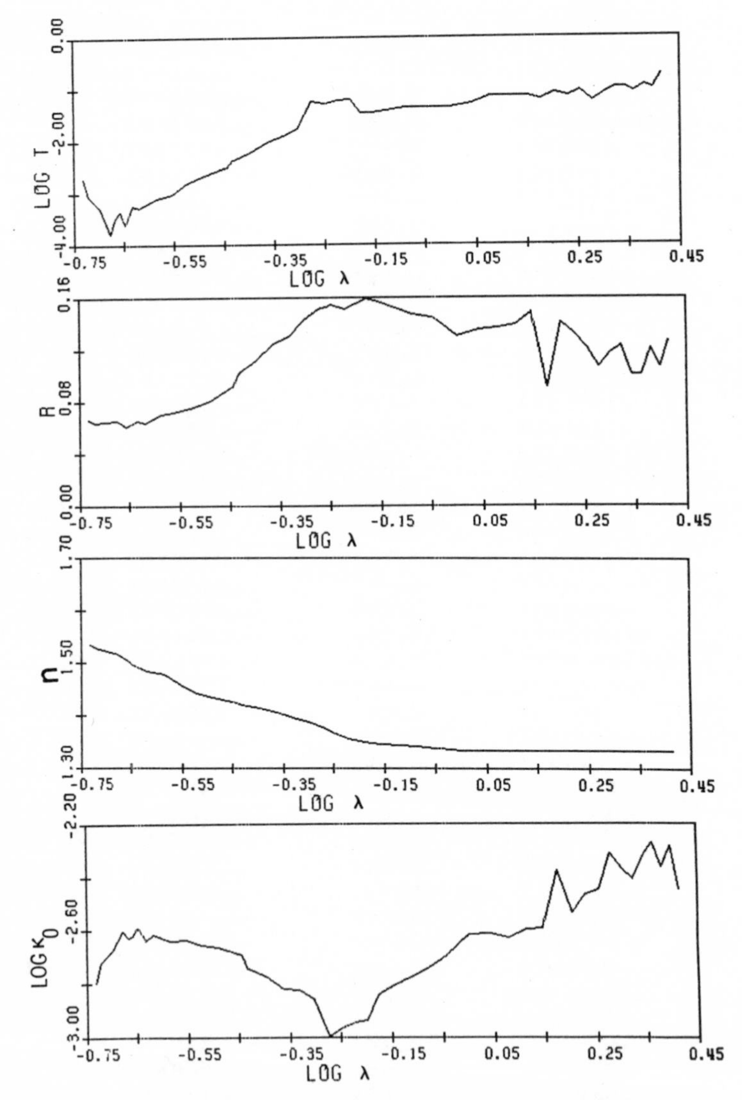

FIG. V-5 Basalt, Columbia River, Bridal Veil Quad, Oregon CP.

TABLE V-7 Basalt, Columbia River, Bridal Veil Quad, Oregon CP[a]

LAMBDA	LOG T	R	LOG K_0	n
0.185	-.272E+01	0.067	-.280E+01	1.533
0.190	-.305E+01	0.064	-.272E+01	1.527
0.200	-.330E+01	0.064	-.267E+01	1.521
0.210	-.380E+01	0.066	-.260E+01	1.515
0.215	-.347E+01	0.064	-.263E+01	1.509
0.220	-.334E+01	0.061	-.262E+01	1.503
0.225	-.360E+01	0.063	-.259E+01	1.496
0.233	-.323E+01	0.066	-.264E+01	1.490
0.240	-.327E+01	0.064	-.261E+01	1.484
0.260	-.310E+01	0.071	-.264E+01	1.478
0.280	-.301E+01	0.072	-.264E+01	1.458
0.300	-.280E+01	0.076	-.266E+01	1.441
0.325	-.266E+01	0.081	-.267E+01	1.433
0.360	-.251E+01	0.092	-.269E+01	1.424
0.370	-.235E+01	0.102	-.275E+01	1.419
0.400	-.222E+01	0.112	-.277E+01	1.413
0.433	-.201E+01	0.125	-.282E+01	1.405
0.466	-.190E+01	0.130	-.282E+01	1.396
0.500	-.176E+01	0.143	-.286E+01	1.388
0.533	-.124E+01	0.151	-.300E+01	1.377
0.566	-.127E+01	0.154	-.296E+01	1.366
0.600	-.122E+01	0.151	-.295E+01	1.355
0.633	-.120E+01	0.155	-.294E+01	1.352
0.666	-.147E+01	0.159	-.284E+01	1.348
0.700	-.146E+01	0.157	-.282E+01	1.345
0.817	-.135E+01	0.147	-.275E+01	1.341
0.907	-.136E+01	0.144	-.270E+01	1.338
1.000	-.136E+01	0.130	-.261E+01	1.334
1.105	-.128E+01	0.135	-.261E+01	1.334
1.200	-.114E+01	0.137	-.263E+01	1.334
1.303	-.115E+01	0.139	-.259E+01	1.334
1.400	-.113E+01	0.149	-.259E+01	1.334
1.500	-.121E+01	0.091	-.237E+01	1.334
1.600	-.109E+01	0.141	-.253E+01	1.334
1.700	-.114E+01	0.132	-.246E+01	1.334
1.800	-.103E+01	0.121	-.245E+01	1.334
1.900	-.124E+01	0.107	-.231E+01	1.334
2.000	-.109E+01	0.118	-.237E+01	1.334
2.100	-.983E+00	0.123	-.241E+01	1.334
2.200	-.987E+00	0.101	-.232E+01	1.334
2.300	-.106E+01	0.101	-.227E+01	1.334
2.400	-.936E+00	0.121	-.236E+01	1.334
2.500	-.991E+00	0.107	-.228E+01	1.334
2.600	-.730E+00	0.127	-.245E+01	1.334

[a] Effective thickness, 0.00414 cm.

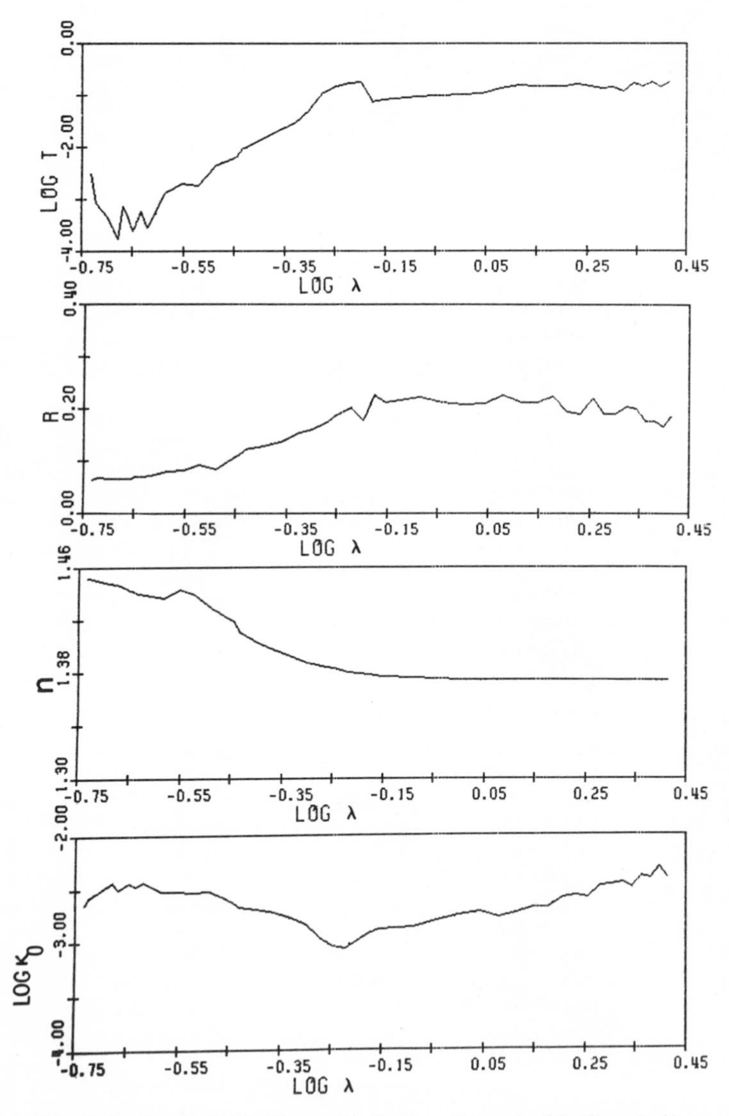

FIG. V-6 Basalt, Table Mountain, Golden, Colorado CP (Sample 1).

TABLE V-8 Basalt, Table Mountain, Golden, Colorado CP (Sample 1)[a]

LAMBDA	LOG T	R	LOG K_0	n
0.185	-.250E+01	0.064	-.265E+01	1.452
0.190	-.307E+01	0.067	-.257E+01	1.451
0.200	-.336E+01	0.067	-.250E+01	1.449
0.210	-.378E+01	0.067	-.243E+01	1.447
0.215	-.312E+01	0.065	-.250E+01	1.446
0.220	-.334E+01	0.067	-.247E+01	1.444
0.225	-.361E+01	0.070	-.244E+01	1.442
0.233	-.322E+01	0.070	-.247E+01	1.440
0.240	-.354E+01	0.071	-.243E+01	1.439
0.260	-.289E+01	0.080	-.252E+01	1.437
0.280	-.271E+01	0.083	-.252E+01	1.443
0.300	-.275E+01	0.093	-.253E+01	1.439
0.325	-.237E+01	0.083	-.252E+01	1.429
0.360	-.217E+01	0.112	-.262E+01	1.418
0.370	-.202E+01	0.122	-.267E+01	1.410
0.400	-.187E+01	0.128	-.269E+01	1.402
0.433	-.168E+01	0.137	-.272E+01	1.397
0.466	-.154E+01	0.152	-.277E+01	1.392
0.500	-.127E+01	0.160	-.283E+01	1.387
0.533	-.939E+00	0.172	-.296E+01	1.385
0.566	-.812E+00	0.190	-.304E+01	1.383
0.600	-.767E+00	0.201	-.306E+01	1.380
0.633	-.742E+00	0.176	-.299E+01	1.379
0.666	-.111E+01	0.226	-.292E+01	1.378
0.700	-.108E+01	0.213	-.288E+01	1.377
0.817	-.102E+01	0.222	-.286E+01	1.376
0.907	-.996E+00	0.212	-.280E+01	1.375
1.000	-.979E+00	0.208	-.275E+01	1.374
1.105	-.959E+00	0.209	-.272E+01	1.374
1.200	-.854E+00	0.225	-.277E+01	1.374
1.303	-.793E+00	0.213	-.273E+01	1.374
1.400	-.830E+00	0.211	-.268E+01	1.374
1.500	-.827E+00	0.223	-.268E+01	1.374
1.600	-.812E+00	0.195	-.259E+01	1.374
1.700	-.772E+00	0.189	-.257E+01	1.374
1.800	-.827E+00	0.221	-.259E+01	1.374
1.900	-.863E+00	0.189	-.248E+01	1.374
2.000	-.846E+00	0.189	-.247E+01	1.374
2.100	-.917E+00	0.205	-.245E+01	1.374
2.200	-.754E+00	0.200	-.250E+01	1.374
2.300	-.812E+00	0.176	-.239E+01	1.374
2.400	-.735E+00	0.176	-.241E+01	1.374
2.500	-.848E+00	0.165	-.231E+01	1.374
2.600	-.730E+00	0.185	-.240E+01	1.374

[a] Effective thickness, 0.00258 cm.

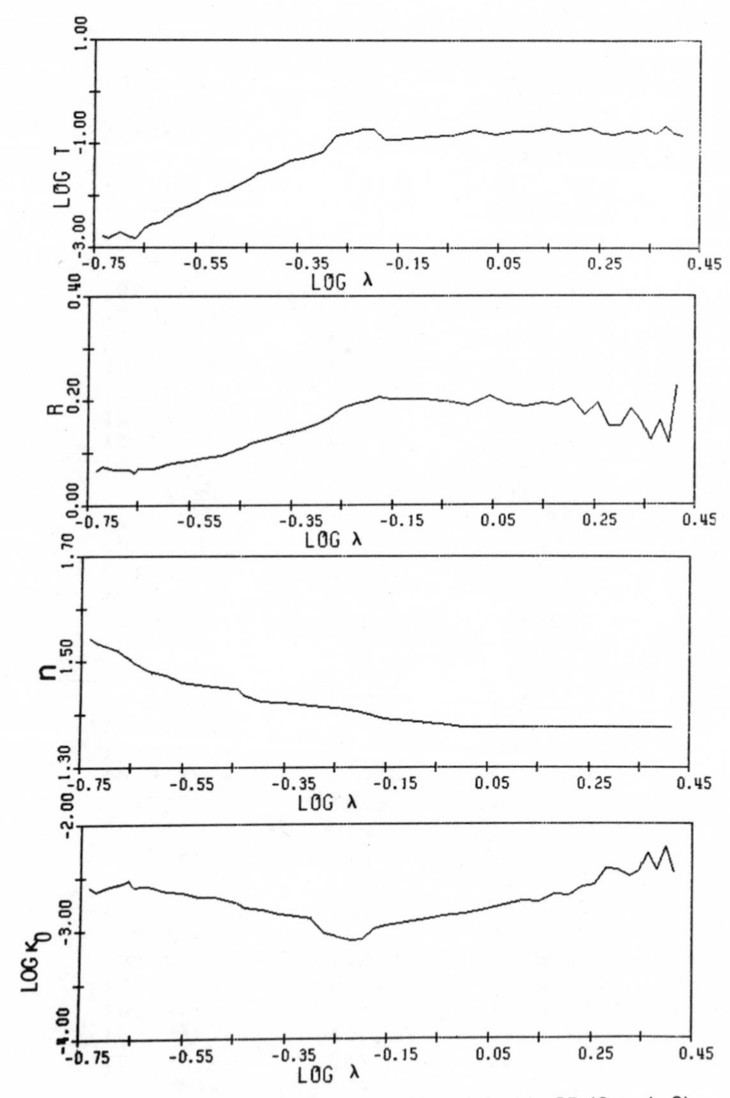

FIG. V-7 Basalt, Table Mountain, Golden, Colorado CP (Sample 2).

TABLE V-9 Basalt, Table Mountain, Golden, Colorado CP (Sample 2)[a]

LAMBDA	LOG T	R	LOG K_0	n
0.185	-.277E+01	0.066	-.259E+01	1.544
0.190	-.281E+01	0.074	-.263E+01	1.536
0.200	-.269E+01	0.068	-.259E+01	1.528
0.210	-.279E+01	0.068	-.256E+01	1.520
0.215	-.281E+01	0.068	-.255E+01	1.512
0.220	-.272E+01	0.061	-.252E+01	1.505
0.225	-.260E+01	0.071	-.259E+01	1.497
0.233	-.255E+01	0.071	-.258E+01	1.489
0.240	-.252E+01	0.071	-.258E+01	1.481
0.260	-.228E+01	0.080	-.263E+01	1.473
0.280	-.216E+01	0.084	-.264E+01	1.459
0.300	-.197E+01	0.090	-.268E+01	1.456
0.325	-.190E+01	0.094	-.268E+01	1.452
0.360	-.169E+01	0.111	-.274E+01	1.448
0.370	-.159E+01	0.118	-.278E+01	1.436
0.400	-.148E+01	0.125	-.280E+01	1.424
0.433	-.133E+01	0.137	-.284E+01	1.421
0.466	-.126E+01	0.145	-.286E+01	1.419
0.500	-.117E+01	0.154	-.288E+01	1.416
0.533	-.848E+00	0.168	-.302E+01	1.413
0.566	-.793E+00	0.185	-.306E+01	1.411
0.600	-.728E+00	0.193	-.309E+01	1.408
0.633	-.717E+00	0.197	-.308E+01	1.403
0.666	-.917E+00	0.206	-.298E+01	1.397
0.700	-.914E+00	0.202	-.295E+01	1.392
0.817	-.854E+00	0.201	-.290E+01	1.386
0.907	-.836E+00	0.197	-.286E+01	1.381
1.000	-.747E+00	0.189	-.284E+01	1.375
1.105	-.818E+00	0.207	-.280E+01	1.375
1.200	-.762E+00	0.191	-.276E+01	1.375
1.303	-.754E+00	0.188	-.272E+01	1.375
1.400	-.710E+00	0.194	-.273E+01	1.375
1.500	-.757E+00	0.189	-.266E+01	1.375
1.600	-.735E+00	0.202	-.267E+01	1.375
1.700	-.703E+00	0.170	-.259E+01	1.375
1.800	-.799E+00	0.195	-.257E+01	1.375
1.900	-.821E+00	0.149	-.242E+01	1.375
2.000	-.763E+00	0.150	-.243E+01	1.375
2.100	-.772E+00	0.182	-.249E+01	1.375
2.200	-.712E+00	0.159	-.244E+01	1.375
2.300	-.796E+00	0.123	-.228E+01	1.375
2.400	-.666E+00	0.161	-.244E+01	1.375
2.500	-.801E+00	0.116	-.222E+01	1.375
2.600	-.848E+00	0.226	-.246E+01	1.375

[a] Effective thickness, 0.00270 cm.

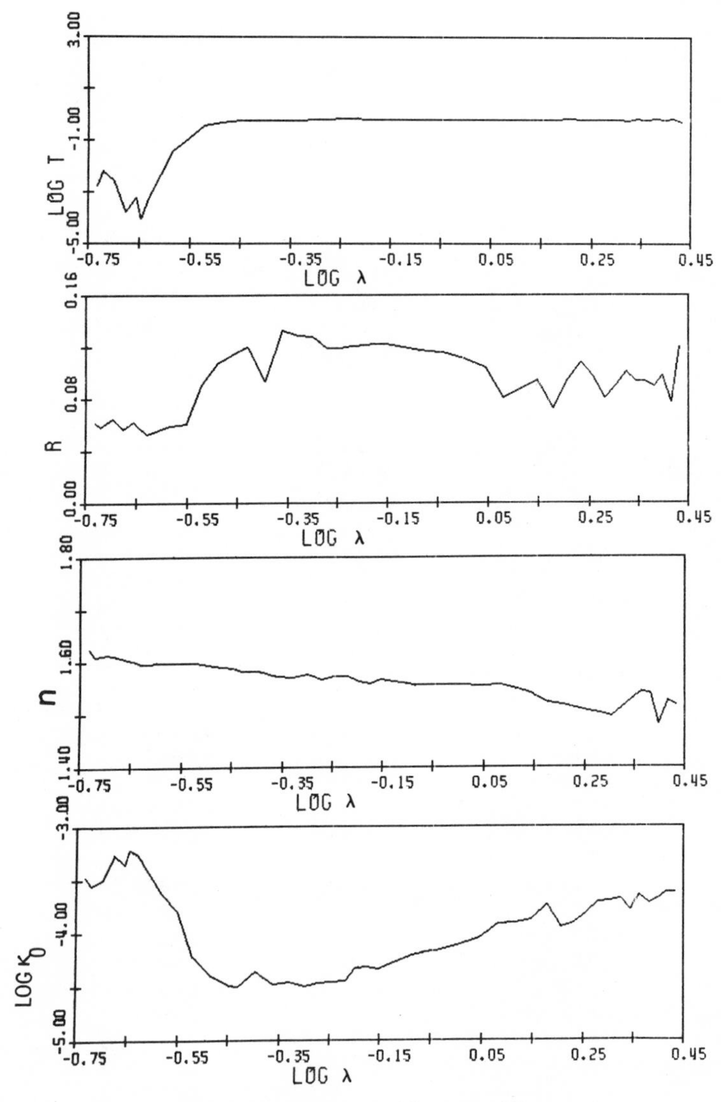

FIG. V-8 Bytownite, Minnesota TS.

TABLE V-10 Bytownite, Minnesota TS[a,b]

LAMBDA	LOG T	R	LOG K_0	n
0.185	-.278E+01	0.061	-.348E+01	1.623
0.190	-.221E+01	0.058	-.356E+01	1.608
0.200	-.264E+01	0.064	-.350E+01	1.612
0.210	-.378E+01	0.057	-.327E+01	1.608
0.220	-.323E+01	0.062	-.336E+01	1.603
0.225	-.408E+01	0.058	-.322E+01	1.600
0.233	-.330E+01	0.053	-.326E+01	1.595
0.260	-.144E+01	0.059	-.363E+01	1.596
0.280	-.937E+00	0.061	-.380E+01	1.597
0.300	-.457E+00	0.090	-.423E+01	1.596
0.325	-.346E+00	0.108	-.439E+01	1.591
0.355	-.275E+00	0.117	-.449E+01	1.587
0.370	-.267E+00	0.119	-.450E+01	1.580
0.400	-.249E+00	0.093	-.437E+01	1.580
0.433	-.271E+00	0.132	-.448E+01	1.571
0.466	-.256E+00	0.128	-.446E+01	1.569
0.500	-.216E+00	0.128	-.451E+01	1.575
0.533	-.203E+00	0.119	-.448E+01	1.564
0.566	-.196E+00	0.119	-.447E+01	1.570
0.600	-.192E+00	0.121	-.446E+01	1.570
0.633	-.240E+00	0.121	-.433E+01	1.561
0.666	-.234E+00	0.122	-.432E+01	1.557
0.700	-.212E+00	0.122	-.435E+01	1.565
0.817	-.231E+00	0.117	-.422E+01	1.555
0.907	-.228E+00	0.116	-.418E+01	1.555
1.000	-.220E+00	0.111	-.413E+01	1.554
1.105	-.216E+00	0.104	-.406E+01	1.553
1.200	-.204E+00	0.080	-.392E+01	1.555
1.303	-.212E+00	0.088	-.392E+01	1.546
1.400	-.231E+00	0.094	-.388E+01	1.537
1.500	-.236E+00	0.072	-.374E+01	1.521
1.600	-.178E+00	0.095	-.396E+01	1.516
1.700	-.212E+00	0.108	-.391E+01	1.510
1.800	-.224E+00	0.097	-.381E+01	1.505
1.900	-.223E+00	0.080	-.371E+01	1.500
2.000	-.240E+00	0.090	-.371E+01	1.495
2.100	-.265E+00	0.100	-.368E+01	1.510
2.200	-.186E+00	0.094	-.379E+01	1.526
2.300	-.236E+00	0.093	-.365E+01	1.541
2.400	-.186E+00	0.089	-.373E+01	1.538
2.500	-.221E+00	0.098	-.368E+01	1.479
2.600	-.190E+00	0.077	-.362E+01	1.524

[a] Effective thickness, 0.0254 cm.

[b] Data taken, in part, from Egan and Hilgeman (1975b).

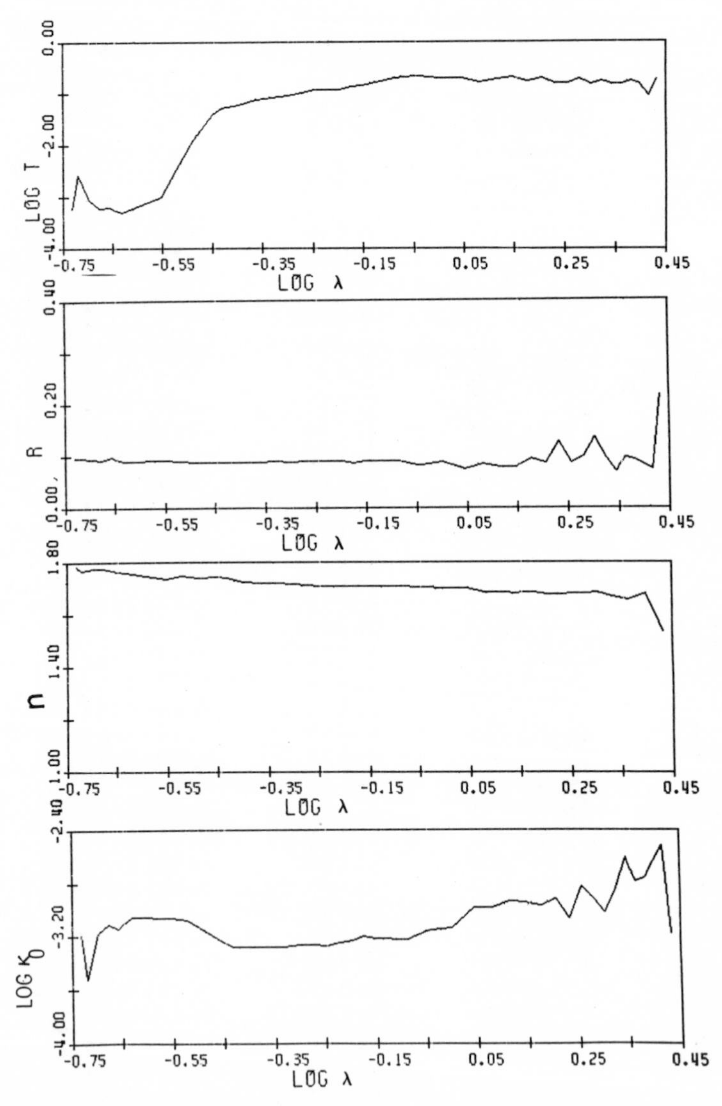

FIG. V-9 Diopside, India TS.

TABLE V-11 Diopside, India TS[a,b]

LAMBDA	LOG T	R	LOG K_0	n
0.185	-.324E+01	0.095	-.319E+01	1.783
0.190	-.257E+01	0.095	-.353E+01	1.767
0.200	-.306E+01	0.093	-.318E+01	1.775
0.210	-.324E+01	0.091	-.312E+01	1.775
0.220	-.322E+01	0.097	-.315E+01	1.767
0.233	-.330E+01	0.089	-.306E+01	1.757
0.280	-.301E+01	0.091	-.307E+01	1.732
0.300	-.248E+01	0.088	-.308E+01	1.746
0.325	-.189E+01	0.086	-.316E+01	1.740
0.355	-.141E+01	0.086	-.325E+01	1.743
0.370	-.129E+01	0.088	-.329E+01	1.739
0.400	-.124E+01	0.088	-.328E+01	1.722
0.433	-.114E+01	0.087	-.329E+01	1.719
0.466	-.109E+01	0.089	-.328E+01	1.717
0.500	-.104E+01	0.088	-.327E+01	1.714
0.533	-.991E+00	0.088	-.327E+01	1.712
0.566	-.939E+00	0.089	-.327E+01	1.707
0.600	-.939E+00	0.090	-.325E+01	1.706
0.633	-.921E+00	0.090	-.324E+01	1.707
0.666	-.879E+00	0.085	-.320E+01	1.707
0.700	-.854E+00	0.089	-.322E+01	1.707
0.817	-.717E+00	0.089	-.322E+01	1.705
0.907	-.666E+00	0.081	-.316E+01	1.703
1.000	-.703E+00	0.088	-.314E+01	1.700
1.105	-.713E+00	0.073	-.299E+01	1.697
1.200	-.799E+00	0.082	-.299E+01	1.681
1.303	-.738E+00	0.076	-.294E+01	1.681
1.400	-.699E+00	0.078	-.294E+01	1.678
1.500	-.775E+00	0.093	-.297E+01	1.680
1.600	-.719E+00	0.084	-.292E+01	1.676
1.700	-.818E+00	0.126	-.307E+01	1.676
1.800	-.804E+00	0.085	-.283E+01	1.678
1.900	-.724E+00	0.097	-.292E+01	1.679
2.000	-.827E+00	0.133	-.303E+01	1.680
2.100	-.775E+00	0.095	-.284E+01	1.671
2.200	-.839E+00	0.067	-.261E+01	1.663
2.300	-.821E+00	0.095	-.279E+01	1.654
2.400	-.775E+00	0.090	-.276E+01	1.666
2.500	-.839E+00	0.081	-.264E+01	1.677
2.600	-.106E+01	0.072	-.252E+01	1.605

[a] Effective thickness, 0.01207 cm.

[b] Data taken, in part, from Egan and Hilgeman (1975b).

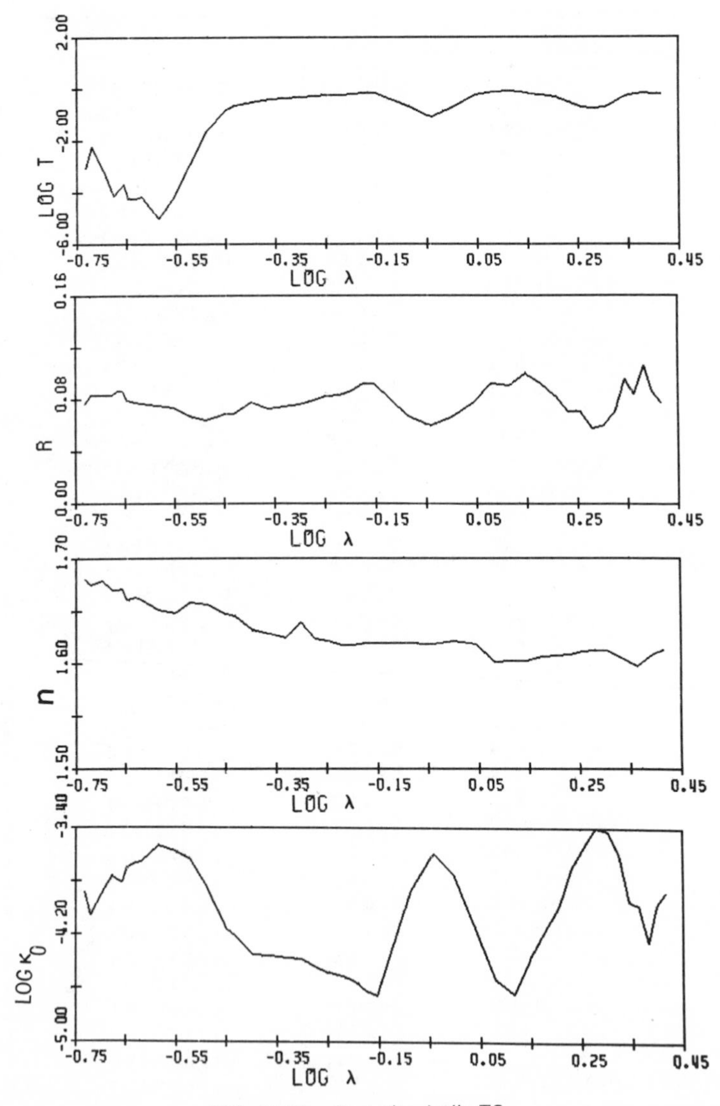

FIG. V-10 Enstatite, India TS.

TABLE V-12 Enstatite, India TS[a,b]

LAMBDA	LOG T	R	LOG K_o	n
0.185	-.309E+01	0.076	-.389E+01	1.680
0.190	-.226E+01	0.083	-.407E+01	1.674
0.200	-.311E+01	0.083	-.390E+01	1.678
0.210	-.415E+01	0.083	-.376E+01	1.669
0.215	-.392E+01	0.086	-.380E+01	1.669
0.220	-.372E+01	0.086	-.381E+01	1.670
0.225	-.424E+01	0.079	-.370E+01	1.660
0.233	-.425E+01	0.077	-.367E+01	1.662
0.240	-.418E+01	0.077	-.366E+01	1.660
0.260	-.503E+01	0.075	-.354E+01	1.650
0.280	-.417E+01	0.073	-.358E+01	1.647
0.300	-.296E+01	0.067	-.364E+01	1.657
0.325	-.166E+01	0.064	-.385E+01	1.655
0.355	-.801E+00	0.068	-.417E+01	1.646
0.370	-.676E+00	0.068	-.422E+01	1.644
0.400	-.530E+00	0.078	-.435E+01	1.631
0.433	-.435E+00	0.072	-.436E+01	1.627
0.466	-.399E+00	0.074	-.438E+01	1.623
0.500	-.356E+00	0.075	-.439E+01	1.638
0.533	-.304E+00	0.079	-.445E+01	1.622
0.566	-.271E+00	0.082	-.449E+01	1.620
0.600	-.247E+00	0.083	-.451E+01	1.616
0.633	-.220E+00	0.086	-.456E+01	1.616
0.666	-.194E+00	0.092	-.462E+01	1.618
0.700	-.170E+00	0.091	-.466E+01	1.618
0.817	-.708E+00	0.067	-.387E+01	1.618
0.907	-.109E+01	0.060	-.360E+01	1.617
1.000	-.724E+00	0.066	-.376E+01	1.620
1.105	-.260E+00	0.077	-.419E+01	1.617
1.200	-.140E+00	0.092	-.454E+01	1.600
1.303	-.108E+00	0.090	-.465E+01	1.601
1.400	-.184E+00	0.100	-.438E+01	1.601
1.500	-.249E+00	0.092	-.417E+01	1.605
1.600	-.321E+00	0.082	-.399E+01	1.606
1.700	-.526E+00	0.070	-.370E+01	1.607
1.800	-.726E+00	0.070	-.354E+01	1.610
1.900	-.777E+00	0.057	-.340E+01	1.611
2.000	-.730E+00	0.060	-.343E+01	1.611
2.100	-.551E+00	0.070	-.359E+01	1.606
2.200	-.299E+00	0.095	-.395E+01	1.601
2.300	-.229E+00	0.083	-.398E+01	1.596
2.400	-.162E+00	0.106	-.425E+01	1.603
2.500	-.213E+00	0.085	-.398E+01	1.609
2.600	-.224E+00	0.077	-.388E+01	1.612

[a] Effective thickness, 0.0635 cm.

[b] Data taken, in part, from Egan and Hilgeman (1975b) and Egan and Hilgeman (1977b).

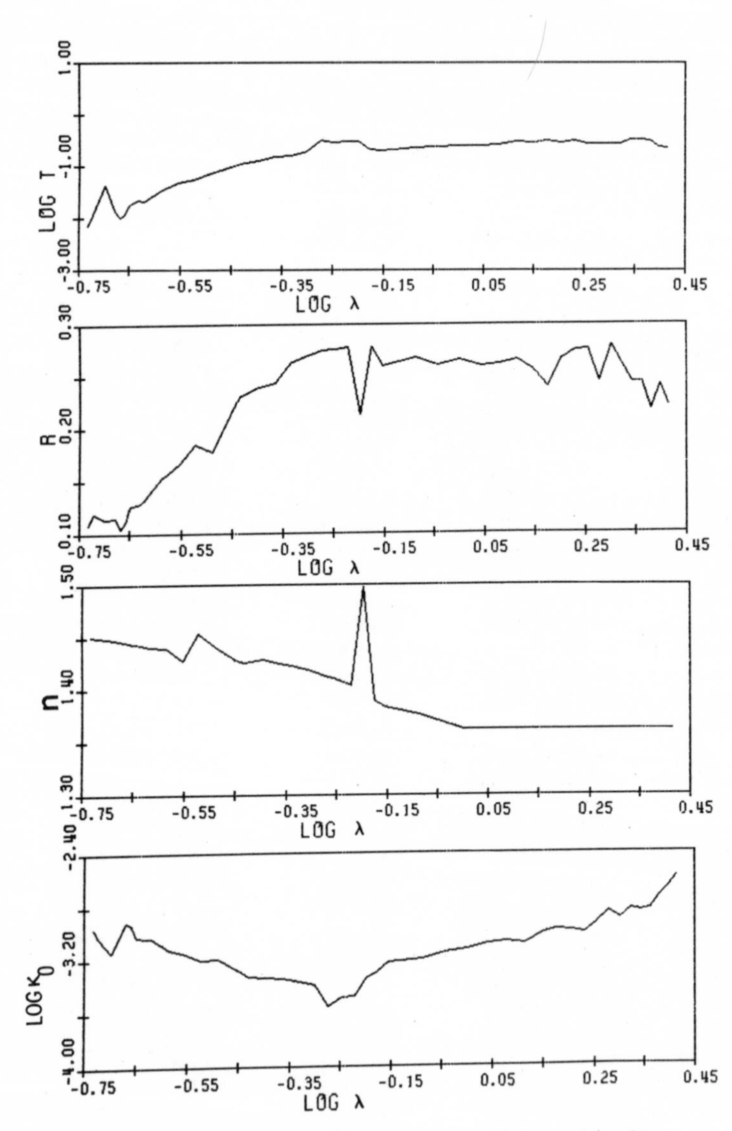

FIG. V-11 Granodiorite, Westerly Granite, Rhode Island CP.

TABLE V-13 Granodiorite, Westerly Granite, Rhode Island CP[a]

LAMBDA	LOG T	R	LOG K_o	n
0.185	-.215E+01	0.109	-.296E+01	1.451
0.190	-.194E+01	0.119	-.303E+01	1.450
0.200	-.136E+01	0.114	-.314E+01	1.449
0.210	-.187E+01	0.115	-.299E+01	1.447
0.215	-.199E+01	0.104	-.291E+01	1.446
0.220	-.193E+01	0.111	-.294E+01	1.445
0.225	-.174E+01	0.127	-.303E+01	1.444
0.233	-.166E+01	0.128	-.303E+01	1.443
0.240	-.167E+01	0.133	-.303E+01	1.441
0.260	-.146E+01	0.154	-.312E+01	1.440
0.280	-.133E+01	0.167	-.316E+01	1.428
0.300	-.127E+01	0.185	-.320E+01	1.454
0.325	-.113E+01	0.178	-.319E+01	1.441
0.360	-.101E+01	0.223	-.330E+01	1.427
0.370	-.951E+00	0.232	-.333E+01	1.426
0.400	-.900E+00	0.240	-.334E+01	1.429
0.433	-.830E+00	0.244	-.334E+01	1.425
0.466	-.796E+00	0.264	-.337E+01	1.422
0.500	-.726E+00	0.270	-.339E+01	1.418
0.533	-.492E+00	0.275	-.355E+01	1.413
0.566	-.535E+00	0.276	-.349E+01	1.409
0.600	-.513E+00	0.278	-.348E+01	1.404
0.633	-.510E+00	0.213	-.334E+01	1.497
0.666	-.686E+00	0.278	-.329E+01	1.390
0.700	-.695E+00	0.260	-.322E+01	1.383
0.817	-.650E+00	0.268	-.320E+01	1.376
0.907	-.627E+00	0.261	-.315E+01	1.369
1.000	-.609E+00	0.266	-.313E+01	1.362
1.105	-.595E+00	0.260	-.309E+01	1.362
1.200	-.577E+00	0.262	-.307E+01	1.362
1.303	-.524E+00	0.266	-.309E+01	1.362
1.400	-.547E+00	0.256	-.301E+01	1.362
1.500	-.499E+00	0.239	-.299E+01	1.362
1.600	-.532E+00	0.267	-.299E+01	1.362
1.700	-.504E+00	0.275	-.301E+01	1.362
1.800	-.558E+00	0.276	-.294E+01	1.362
1.900	-.554E+00	0.244	-.285E+01	1.362
2.000	-.554E+00	0.279	-.290E+01	1.362
2.100	-.564E+00	0.262	-.284E+01	1.362
2.200	-.484E+00	0.244	-.285E+01	1.362
2.300	-.479E+00	0.244	-.283E+01	1.362
2.400	-.493E+00	0.217	-.274E+01	1.362
2.500	-.623E+00	0.241	-.267E+01	1.362
2.600	-.650E+00	0.221	-.259E+01	1.362

[a] Effective thickness, 0.00294 cm.

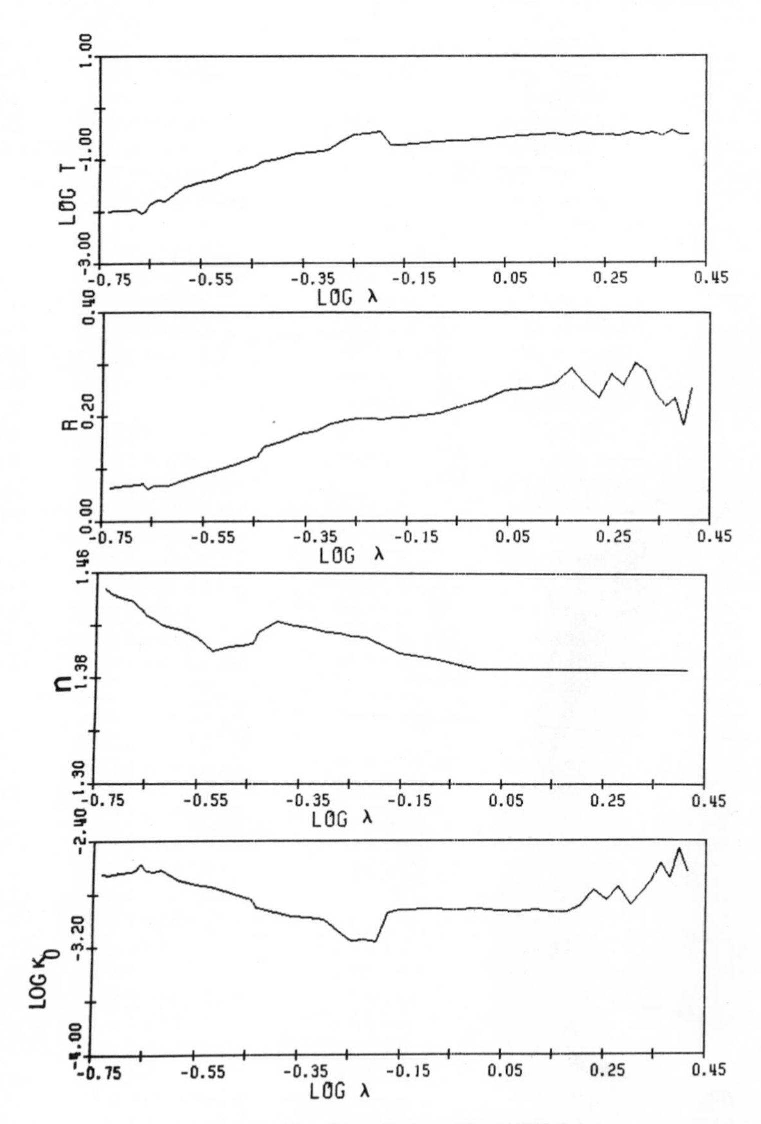

FIG. V-12 Illite, Fithian, Illinois CP.

TABLE V-14 Illite, Fithian, Illinois CP[a]

LAMBDA	LOG T	R	LOG K_0	n
0.185	-.200E+01	0.063	-.265E+01	1.448
0.190	-.198E+01	0.067	-.266E+01	1.444
0.200	-.198E+01	0.068	-.264E+01	1.441
0.210	-.196E+01	0.070	-.263E+01	1.438
0.215	-.204E+01	0.072	-.262E+01	1.434
0.220	-.197E+01	0.061	-.257E+01	1.431
0.225	-.185E+01	0.067	-.262E+01	1.427
0.233	-.177E+01	0.069	-.263E+01	1.424
0.240	-.179E+01	0.068	-.261E+01	1.420
0.260	-.154E+01	0.081	-.269E+01	1.417
0.280	-.144E+01	0.090	-.273E+01	1.411
0.300	-.138E+01	0.098	-.274E+01	1.401
0.325	-.124E+01	0.108	-.278E+01	1.404
0.360	-.111E+01	0.124	-.284E+01	1.406
0.370	-.104E+01	0.141	-.291E+01	1.415
0.400	-.975E+00	0.152	-.293E+01	1.423
0.433	-.889E+00	0.165	-.297E+01	1.420
0.466	-.854E+00	0.172	-.297E+01	1.418
0.500	-.815E+00	0.185	-.299E+01	1.415
0.533	-.644E+00	0.191	-.308E+01	1.414
0.566	-.526E+00	0.196	-.315E+01	1.412
0.600	-.503E+00	0.196	-.315E+01	1.411
0.633	-.456E+00	0.194	-.316E+01	1.407
0.666	-.728E+00	0.197	-.294E+01	1.403
0.700	-.719E+00	0.197	-.292E+01	1.399
0.817	-.658E+00	0.206	-.291E+01	1.395
0.907	-.633E+00	0.221	-.292E+01	1.391
1.000	-.611E+00	0.231	-.291E+01	1.387
1.105	-.588E+00	0.250	-.293E+01	1.387
1.200	-.547E+00	0.253	-.293E+01	1.387
1.303	-.529E+00	0.256	-.292E+01	1.387
1.400	-.504E+00	0.265	-.293E+01	1.387
1.500	-.535E+00	0.293	-.294E+01	1.387
1.600	-.480E+00	0.260	-.289E+01	1.387
1.700	-.514E+00	0.237	-.277E+01	1.387
1.800	-.523E+00	0.282	-.284E+01	1.387
1.900	-.547E+00	0.260	-.275E+01	1.387
2.000	-.486E+00	0.303	-.289E+01	1.387
2.100	-.519E+00	0.288	-.280E+01	1.387
2.200	-.483E+00	0.243	-.271E+01	1.387
2.300	-.547E+00	0.220	-.257E+01	1.387
2.400	-.450E+00	0.236	-.269E+01	1.387
2.500	-.529E+00	0.183	-.246E+01	1.387
2.600	-.510E+00	0.254	-.263E+01	1.387

[a] Effective thickness, 0.00207 cm.

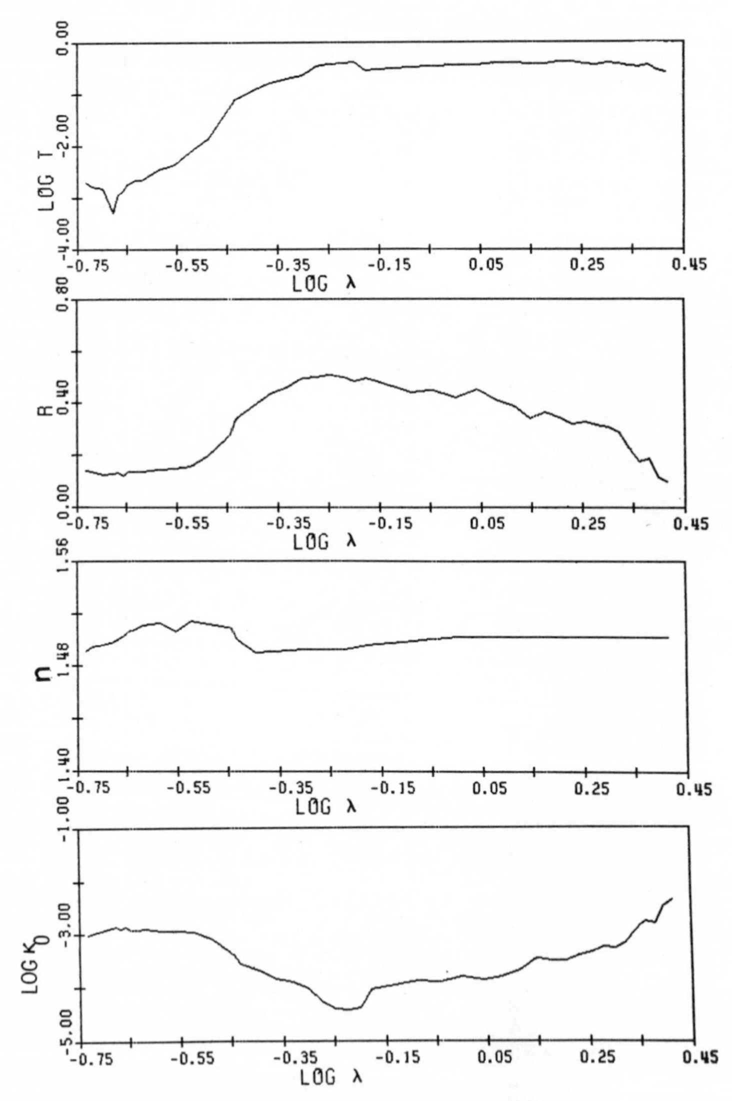

FIG. V-13 Kaolinite, Macon, Georgia CP.

TABLE V-15 Kaolinite, Macon, Georgia CP[a]

LAMBDA	LOG T	R	LOG K_o	n
0.185	-.269E+01	0.139	-.302E+01	1.491
0.190	-.279E+01	0.135	-.298E+01	1.494
0.200	-.283E+01	0.126	-.292E+01	1.496
0.210	-.329E+01	0.129	-.285E+01	1.498
0.215	-.294E+01	0.132	-.290E+01	1.501
0.220	-.287E+01	0.122	-.286E+01	1.503
0.225	-.274E+01	0.134	-.291E+01	1.506
0.233	-.266E+01	0.135	-.292E+01	1.508
0.240	-.266E+01	0.136	-.291E+01	1.511
0.260	-.246E+01	0.145	-.293E+01	1.513
0.280	-.235E+01	0.149	-.293E+01	1.506
0.300	-.211E+01	0.155	-.297E+01	1.514
0.325	-.189E+01	0.195	-.309E+01	1.512
0.360	-.126E+01	0.278	-.340E+01	1.509
0.370	-.110E+01	0.339	-.356E+01	1.500
0.400	-.928E+00	0.387	-.369E+01	1.490
0.433	-.775E+00	0.436	-.384E+01	1.491
0.466	-.706E+00	0.459	-.391E+01	1.492
0.500	-.642E+00	0.497	-.402E+01	1.493
0.533	-.466E+00	0.500	-.428E+01	1.493
0.566	-.425E+00	0.509	-.441E+01	1.493
0.600	-.406E+00	0.501	-.442E+01	1.493
0.633	-.389E+00	0.485	-.438E+01	1.494
0.666	-.535E+00	0.494	-.403E+01	1.496
0.700	-.524E+00	0.485	-.400E+01	1.497
0.817	-.480E+00	0.442	-.389E+01	1.499
0.907	-.457E+00	0.449	-.391E+01	1.501
1.000	-.446E+00	0.421	-.380E+01	1.502
1.105	-.439E+00	0.452	-.387E+01	1.502
1.200	-.393E+00	0.411	-.380E+01	1.502
1.303	-.395E+00	0.387	-.369E+01	1.502
1.400	-.429E+00	0.340	-.346E+01	1.502
1.500	-.420E+00	0.363	-.351E+01	1.502
1.600	-.378E+00	0.344	-.350E+01	1.502
1.700	-.385E+00	0.319	-.339E+01	1.502
1.800	-.412E+00	0.327	-.334E+01	1.502
1.900	-.444E+00	0.317	-.325E+01	1.502
2.000	-.409E+00	0.309	-.325E+01	1.502
2.100	-.428E+00	0.290	-.316E+01	1.502
2.200	-.469E+00	0.226	-.292E+01	1.502
2.300	-.480E+00	0.176	-.275E+01	1.502
2.400	-.442E+00	0.187	-.280E+01	1.502
2.500	-.550E+00	0.118	-.248E+01	1.502
2.600	-.580E+00	0.100	-.237E+01	1.502

[a] Effective thickness, 0.00327 cm.

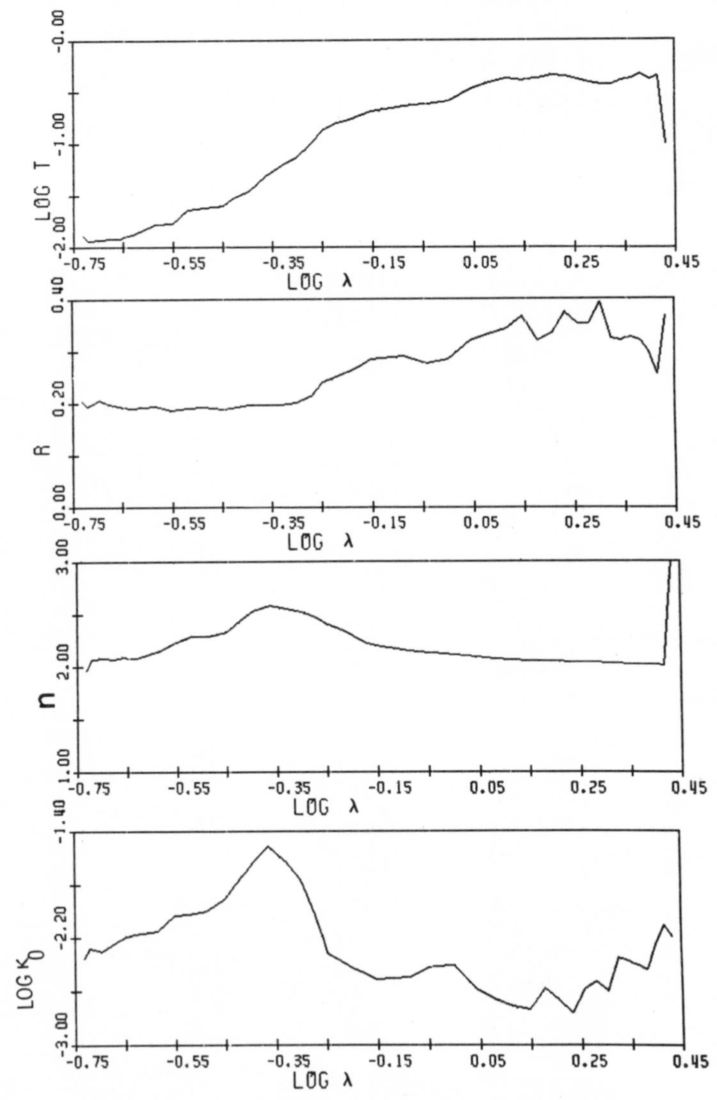

FIG. V-14 Limonite, Venango County, Pennsylvania CP.

TABLE V-16 Limonite, Venango County, Pennsylvania CP[a]

LAMBDA	LOG T	R	LOG K_o	n
0.185	-.190E+01	0.203	-.236E+01	1.961
0.190	-.195E+01	0.194	-.228E+01	2.070
0.200	-.194E+01	0.204	-.231E+01	2.077
0.210	-.193E+01	0.197	-.226E+01	2.063
0.220	-.193E+01	0.193	-.222E+01	2.089
0.233	-.188E+01	0.189	-.218E+01	2.078
0.260	-.178E+01	0.192	-.215E+01	2.150
0.280	-.178E+01	0.186	-.204E+01	2.223
0.300	-.165E+01	0.189	-.202E+01	2.282
0.325	-.163E+01	0.190	-.201E+01	2.281
0.355	-.161E+01	0.188	-.191E+01	2.330
0.370	-.153E+01	0.189	-.181E+01	2.410
0.400	-.146E+01	0.195	-.165E+01	2.528
0.433	-.131E+01	0.196	-.152E+01	2.579
0.466	-.122E+01	0.195	-.162E+01	2.544
0.500	-.113E+01	0.198	-.177E+01	2.513
0.533	-.100E+01	0.211	-.203E+01	2.464
0.566	-.860E+00	0.240	-.232E+01	2.400
0.600	-.810E+00	0.249	-.238E+01	2.350
0.633	-.772E+00	0.260	-.243E+01	2.285
0.666	-.735E+00	0.269	-.247E+01	2.228
0.700	-.697E+00	0.283	-.251E+01	2.200
0.817	-.640E+00	0.290	-.250E+01	2.146
0.907	-.622E+00	0.276	-.242E+01	2.122
1.000	-.593E+00	0.283	-.242E+01	2.108
1.105	-.471E+00	0.320	-.260E+01	2.082
1.200	-.412E+00	0.331	-.267E+01	2.069
1.300	-.379E+00	0.341	-.272E+01	2.055
1.400	-.396E+00	0.366	-.274E+01	2.050
1.500	-.373E+00	0.319	-.258E+01	2.045
1.600	-.340E+00	0.333	-.267E+01	2.041
1.700	-.356E+00	0.373	-.276E+01	2.038
1.800	-.389E+00	0.351	-.259E+01	2.035
1.900	-.412E+00	0.352	-.253E+01	2.033
2.000	-.432E+00	0.392	-.261E+01	2.030
2.100	-.437E+00	0.323	-.236E+01	2.025
2.200	-.394E+00	0.320	-.238E+01	2.019
2.300	-.376E+00	0.326	-.241E+01	2.017
2.400	-.336E+00	0.319	-.244E+01	2.014
2.500	-.382E+00	0.298	-.227E+01	2.012
2.600	-.360E+00	0.254	-.211E+01	2.004

[a] Effective thickness, 0.00029 cm.

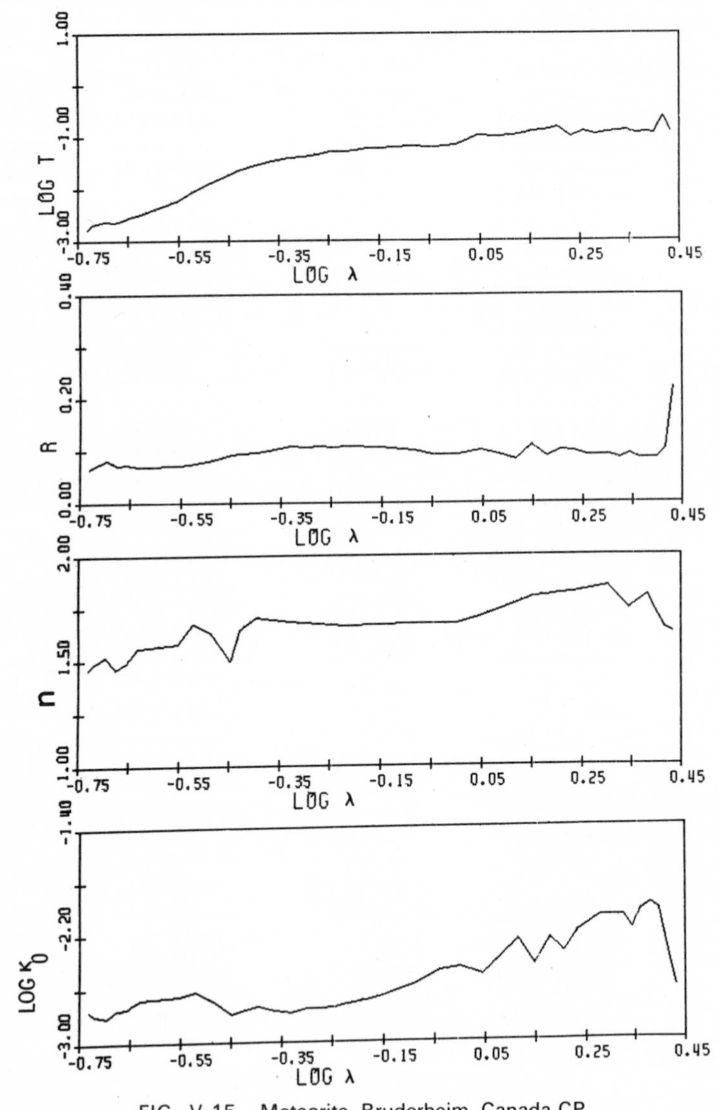

FIG. V-15 Meteorite, Bruderheim, Canada CP.

TABLE V-17 Meteorite, Bruderheim, Canada CP[a,b]

LAMBDA	LOG T	R	LOG K_0	n
0.185	-.278E+01	0.064	-.276E+01	1.459
0.190	-.269E+01	0.072	-.279E+01	1.485
0.200	-.263E+01	0.081	-.281E+01	1.519
0.210	-.264E+01	0.070	-.276E+01	1.458
0.220	-.259E+01	0.072	-.275E+01	1.487
0.233	-.250E+01	0.069	-.269E+01	1.559
0.280	-.223E+01	0.071	-.266E+01	1.579
0.300	-.204E+01	0.075	-.263E+01	1.673
0.325	-.190E+01	0.081	-.269E+01	1.633
0.355	-.174E+01	0.091	-.280E+01	1.495
0.370	-.164E+01	0.094	-.277E+01	1.642
0.400	-.155E+01	0.095	-.274E+01	1.701
0.433	-.147E+01	0.101	-.277E+01	1.693
0.466	-.141E+01	0.105	-.280E+01	1.684
0.500	-.138E+01	0.107	-.275E+01	1.676
0.533	-.135E+01	0.114	-.272E+01	1.671
0.566	-.129E+01	0.106	-.275E+01	1.666
0.600	-.129E+01	0.106	-.273E+01	1.661
0.633	-.128E+01	0.106	-.271E+01	1.664
0.666	-.123E+01	0.106	-.270E+01	1.666
0.700	-.123E+01	0.105	-.267E+01	1.668
0.817	-.120E+01	0.099	-.258E+01	1.671
0.907	-.120E+01	0.091	-.248E+01	1.672
1.000	-.116E+01	0.091	-.246E+01	1.675
1.105	-.996E+00	0.100	-.252E+01	1.705
1.200	-.101E+01	0.090	-.239E+01	1.734
1.303	-.983E+00	0.081	-.226E+01	1.764
1.400	-.924E+00	0.109	-.245E+01	1.793
1.500	-.886E+00	0.086	-.225E+01	1.802
1.600	-.839E+00	0.100	-.235E+01	1.810
1.700	-.102E+01	0.097	-.221E+01	1.819
1.800	-.917E+00	0.089	-.214E+01	1.828
1.900	-.979E+00	0.089	-.208E+01	1.836
2.000	-.924E+00	0.090	-.208E+01	1.845
2.100	-.901E+00	0.084	-.208E+01	1.792
2.200	-.889E+00	0.091	-.218E+01	1.739
2.300	-.955E+00	0.084	-.204E+01	1.772
2.400	-.924E+00	0.084	-.200E+01	1.804
2.500	-.947E+00	0.083	-.205E+01	1.729
2.600	-.620E+00	0.100	-.236E+01	1.654

[a] Effective thickness, 0.00327 cm.

[b] Data taken, in part, from Egan and Hilgeman (1975b).

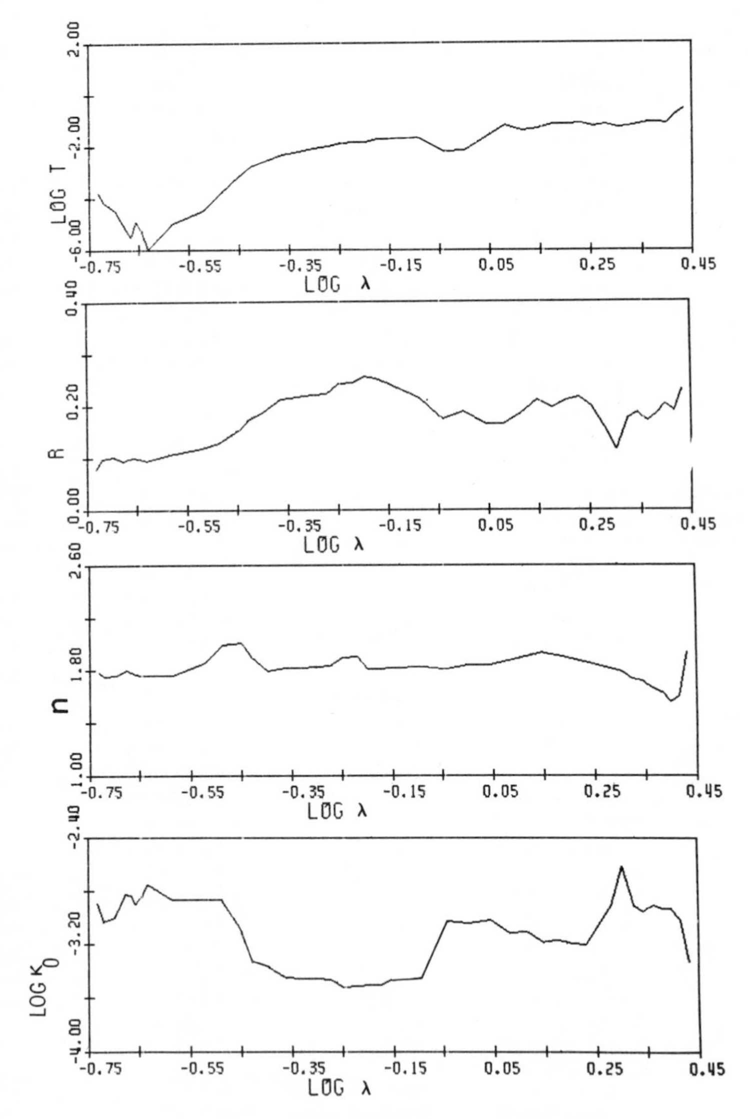

FIG. V-16 Meteorite, Bruderheim, Canada TS.

TABLE V-18 Meteorite, Bruderheim, Canada TS[a,b]

LAMBDA	LOG T	R	LOG K_0	n
0.185	-.380E+01	0.079	-.290E+01	1.780
0.190	-.419E+01	0.096	-.303E+01	1.747
0.200	-.449E+01	0.100	-.300E+01	1.753
0.210	-.520E+01	0.093	-.283E+01	1.794
0.215	-.551E+01	0.096	-.283E+01	1.779
0.220	-.493E+01	0.099	-.291E+01	1.764
0.225	-.522E+01	0.096	-.285E+01	1.760
0.233	-.600E+01	0.093	-.275E+01	1.758
0.260	-.504E+01	0.104	-.287E+01	1.756
0.300	-.449E+01	0.117	-.287E+01	1.856
0.325	-.381E+01	0.127	-.286E+01	1.986
0.355	-.315E+01	0.150	-.310E+01	2.002
0.370	-.285E+01	0.168	-.328E+01	1.902
0.400	-.259E+01	0.187	-.337E+01	1.792
0.433	-.234E+01	0.210	-.345E+01	1.812
0.466	-.223E+01	0.214	-.346E+01	1.812
0.500	-.212E+01	0.219	-.346E+01	1.819
0.533	-.200E+01	0.220	-.346E+01	1.825
0.566	-.189E+01	0.240	-.353E+01	1.889
0.600	-.187E+01	0.242	-.351E+01	1.904
0.633	-.187E+01	0.255	-.351E+01	1.803
0.666	-.174E+01	0.250	-.351E+01	1.808
0.700	-.173E+01	0.242	-.347E+01	1.813
0.817	-.193E+01	0.212	-.327E+01	1.818
0.907	-.222E+01	0.173	-.303E+01	1.803
1.000	-.219E+01	0.187	-.305E+01	1.834
1.105	-.167E+01	0.151	-.296E+01	1.834
1.200	-.121E+01	0.163	-.312E+01	1.866
1.303	-.141E+01	0.183	-.310E+01	1.898
1.400	-.134E+01	0.208	-.319E+01	1.930
1.500	-.118E+01	0.195	-.317E+01	1.907
1.600	-.117E+01	0.209	-.319E+01	1.883
1.700	-.112E+01	0.214	-.320E+01	1.860
1.800	-.126E+01	0.194	-.305E+01	1.836
1.900	-.116E+01	0.156	-.291E+01	1.813
2.000	-.130E+01	0.113	-.261E+01	1.789
2.100	-.126E+01	0.173	-.291E+01	1.736
2.200	-.118E+01	0.185	-.296E+01	1.717
2.300	-.109E+01	0.168	-.291E+01	1.664
2.400	-.109E+01	0.183	-.293E+01	1.629
2.500	-.113E+01	0.201	-.293E+01	1.558
2.600	-.827E+00	0.188	-.302E+01	1.598

[a] Effective thickness, 0.01 cm.

[b] Data taken, in part, from Egan and Hilgeman (1975b).

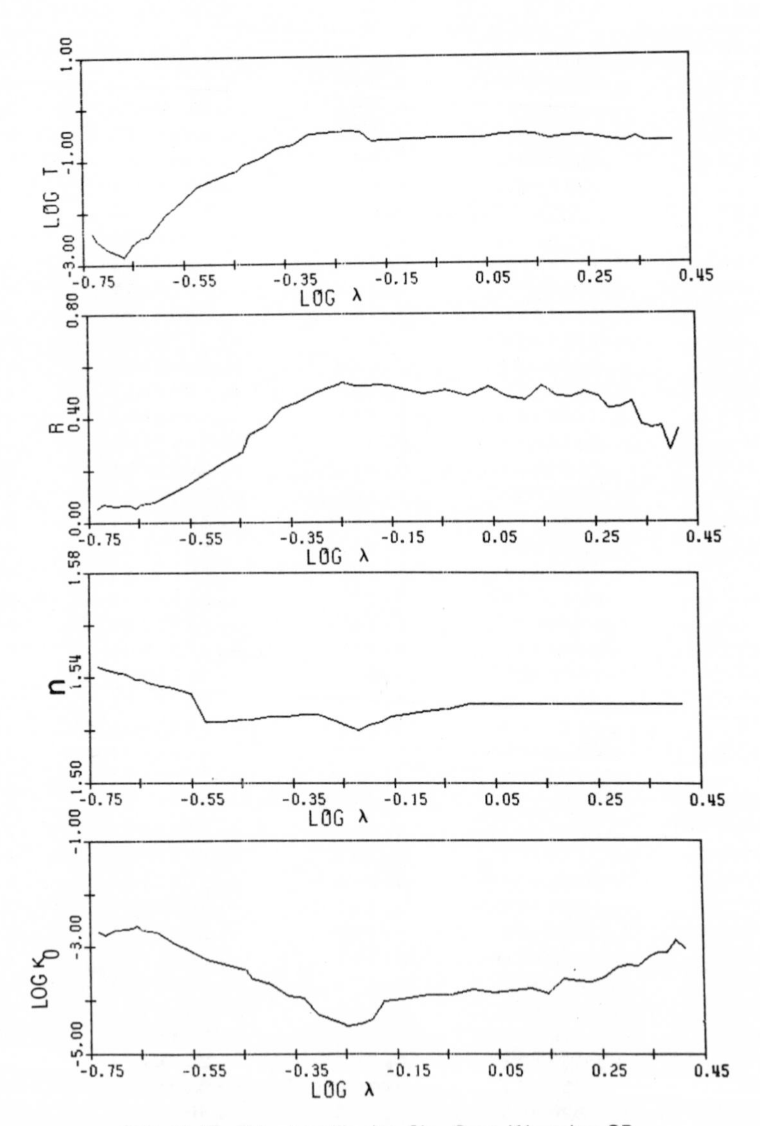

FIG. V-17 Montmorillonite, Clay Spur, Wyoming CP.

TABLE V-19 Montmorillonite, Clay Spur, Wyoming CP[a]

LAMBDA	LOG T	R	LOG K_o	n
0.185	-.238E+01	0.056	-.272E+01	1.544
0.190	-.256E+01	0.074	-.278E+01	1.543
0.200	-.273E+01	0.066	-.268E+01	1.542
0.210	-.281E+01	0.067	-.266E+01	1.541
0.215	-.286E+01	0.068	-.265E+01	1.540
0.220	-.273E+01	0.057	-.259E+01	1.539
0.225	-.259E+01	0.071	-.269E+01	1.539
0.233	-.249E+01	0.074	-.271E+01	1.538
0.240	-.246E+01	0.078	-.272E+01	1.537
0.260	-.204E+01	0.115	-.293E+01	1.536
0.280	-.177E+01	0.147	-.308E+01	1.534
0.300	-.150E+01	0.186	-.323E+01	1.523
0.325	-.137E+01	0.224	-.333E+01	1.523
0.360	-.120E+01	0.273	-.345E+01	1.524
0.370	-.108E+01	0.335	-.360E+01	1.524
0.400	-.959E+00	0.371	-.369E+01	1.525
0.433	-.757E+00	0.442	-.391E+01	1.525
0.466	-.706E+00	0.464	-.397E+01	1.526
0.500	-.496E+00	0.497	-.428E+01	1.526
0.533	-.472E+00	0.517	-.437E+01	1.524
0.566	-.453E+00	0.534	-.447E+01	1.522
0.600	-.441E+00	0.523	-.444E+01	1.520
0.633	-.461E+00	0.523	-.435E+01	1.522
0.666	-.636E+00	0.528	-.403E+01	1.523
0.700	-.627E+00	0.525	-.401E+01	1.525
0.817	-.599E+00	0.491	-.389E+01	1.527
0.907	-.587E+00	0.506	-.390E+01	1.528
1.000	-.587E+00	0.482	-.380E+01	1.530
1.105	-.577E+00	0.519	-.387E+01	1.530
1.200	-.519E+00	0.479	-.381E+01	1.530
1.303	-.495E+00	0.467	-.378E+01	1.530
1.400	-.520E+00	0.525	-.388E+01	1.530
1.500	-.606E+00	0.485	-.361E+01	1.530
1.600	-.554E+00	0.480	-.363E+01	1.530
1.700	-.550E+00	0.498	-.366E+01	1.530
1.800	-.562E+00	0.485	-.358E+01	1.530
1.900	-.600E+00	0.435	-.339E+01	1.530
2.000	-.639E+00	0.438	-.334E+01	1.530
2.100	-.658E+00	0.464	-.336E+01	1.530
2.200	-.580E+00	0.371	-.321E+01	1.530
2.300	-.656E+00	0.360	-.310E+01	1.530
2.400	-.654E+00	0.369	-.310E+01	1.530
2.500	-.670E+00	0.273	-.286E+01	1.530
2.600	-.666E+00	0.352	-.302E+01	1.530

[a] Effective thickness, 0.00352 cm.

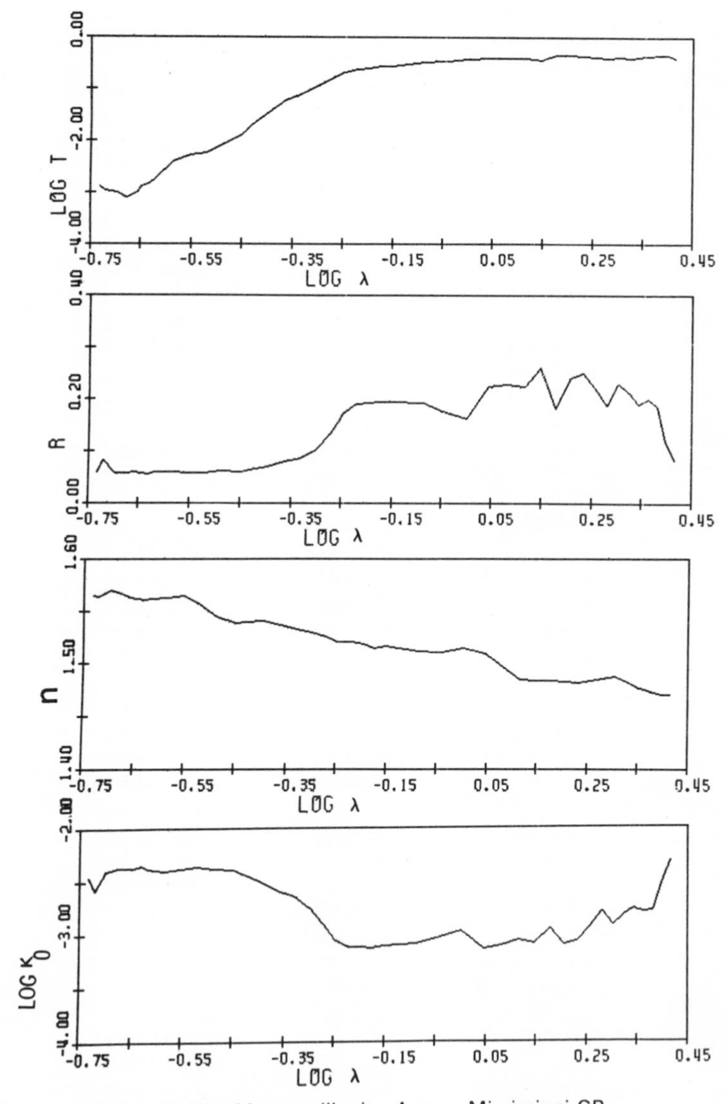

FIG. V-18 Montmorillonite, Amory, Mississippi CP.

TABLE V-20 Montmorillonite, Amory, Mississippi CP[a]

LAMBDA	LOG T	R	LOG K_0	n
0.185	-.290E+01	0.058	-.246E+01	1.564
0.190	-.296E+01	0.082	-.259E+01	1.563
0.200	-.302E+01	0.058	-.240E+01	1.569
0.210	-.312E+01	0.058	-.237E+01	1.566
0.220	-.301E+01	0.059	-.238E+01	1.562
0.225	-.288E+01	0.058	-.238E+01	1.561
0.233	-.282E+01	0.055	-.236E+01	1.560
0.240	-.274E+01	0.059	-.238E+01	1.561
0.260	-.241E+01	0.059	-.241E+01	1.562
0.280	-.229E+01	0.058	-.238E+01	1.564
0.300	-.226E+01	0.058	-.236E+01	1.556
0.325	-.209E+01	0.061	-.239E+01	1.544
0.355	-.189E+01	0.060	-.240E+01	1.538
0.370	-.171E+01	0.062	-.243E+01	1.539
0.400	-.147E+01	0.068	-.250E+01	1.540
0.433	-.123E+01	0.078	-.260E+01	1.536
0.466	-.111E+01	0.085	-.265E+01	1.532
0.500	-.955E+00	0.102	-.275E+01	1.529
0.533	-.804E+00	0.133	-.291E+01	1.525
0.566	-.689E+00	0.171	-.307E+01	1.520
0.600	-.639E+00	0.189	-.312E+01	1.520
0.633	-.614E+00	0.192	-.313E+01	1.518
0.666	-.575E+00	0.194	-.313E+01	1.514
0.700	-.565E+00	0.193	-.312E+01	1.516
0.817	-.498E+00	0.191	-.310E+01	1.511
0.907	-.468E+00	0.174	-.303E+01	1.510
1.000	-.438E+00	0.161	-.298E+01	1.514
1.105	-.409E+00	0.223	-.315E+01	1.509
1.200	-.416E+00	0.227	-.311E+01	1.496
1.303	-.410E+00	0.223	-.307E+01	1.484
1.400	-.446E+00	0.260	-.309E+01	1.483
1.500	-.358E+00	0.181	-.295E+01	1.483
1.600	-.349E+00	0.239	-.311E+01	1.482
1.700	-.374E+00	0.250	-.307E+01	1.481
1.800	-.393E+00	0.221	-.294E+01	1.483
1.900	-.419E+00	0.187	-.279E+01	1.485
2.000	-.394E+00	0.229	-.292E+01	1.487
2.100	-.409E+00	0.212	-.283E+01	1.483
2.200	-.386E+00	0.189	-.277E+01	1.478
2.300	-.376E+00	0.199	-.279E+01	1.475
2.400	-.340E+00	0.185	-.278E+01	1.472
2.500	-.346E+00	0.120	-.254E+01	1.470
2.600	-.401E+00	0.081	-.232E+01	1.470

[a] Effective thickness, 0.0024 cm.

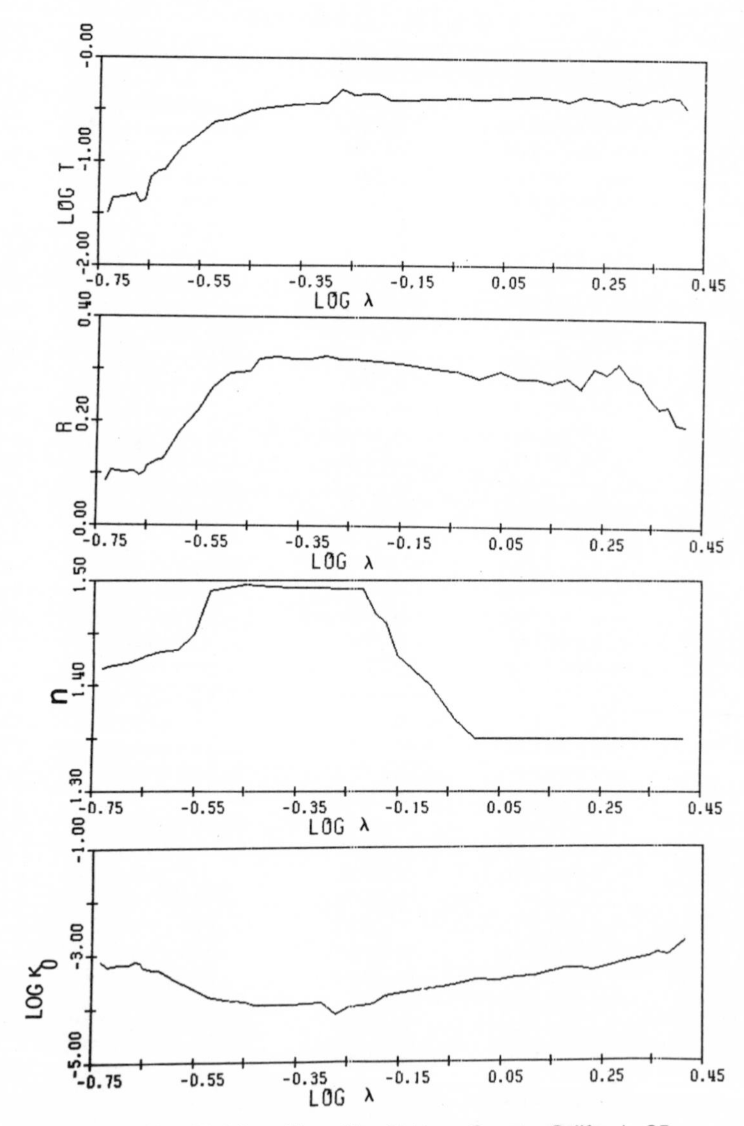

FIG. V-19 Obsidian, Glass Mt., Siskiyou County, California CP.

TABLE V-21 Obsidian, Glass Mt., Siskiyou County, California CP[a]

LAMBDA	LOG T	R	LOG K_o	n
0.185	-.149E+01	0.086	-.311E+01	1.416
0.190	-.135E+01	0.106	-.322E+01	1.418
0.200	-.134E+01	0.101	-.318E+01	1.420
0.210	-.131E+01	0.104	-.318E+01	1.422
0.215	-.139E+01	0.095	-.311E+01	1.424
0.220	-.136E+01	0.102	-.314E+01	1.426
0.225	-.115E+01	0.116	-.324E+01	1.428
0.233	-.109E+01	0.123	-.328E+01	1.430
0.240	-.108E+01	0.128	-.328E+01	1.432
0.260	-.857E+00	0.180	-.349E+01	1.434
0.280	-.742E+00	0.219	-.362E+01	1.449
0.300	-.622E+00	0.264	-.379E+01	1.490
0.325	-.587E+00	0.291	-.384E+01	1.493
0.355	-.514E+00	0.295	-.388E+01	1.496
0.370	-.492E+00	0.319	-.394E+01	1.495
0.400	-.470E+00	0.323	-.394E+01	1.494
0.433	-.449E+00	0.321	-.393E+01	1.493
0.466	-.438E+00	0.320	-.391E+01	1.493
0.500	-.428E+00	0.326	-.391E+01	1.493
0.533	-.303E+00	0.320	-.411E+01	1.492
0.566	-.351E+00	0.321	-.397E+01	1.492
0.600	-.340E+00	0.317	-.396E+01	1.492
0.633	-.339E+00	0.315	-.392E+01	1.470
0.666	-.399E+00	0.313	-.378E+01	1.460
0.700	-.397E+00	0.311	-.374E+01	1.430
0.817	-.385E+00	0.301	-.365E+01	1.400
0.907	-.384E+00	0.298	-.358E+01	1.370
1.000	-.393E+00	0.284	-.348E+01	1.350
1.105	-.380E+00	0.298	-.349E+01	1.350
1.200	-.371E+00	0.284	-.343E+01	1.350
1.303	-.360E+00	0.283	-.341E+01	1.350
1.400	-.376E+00	0.276	-.334E+01	1.350
1.500	-.413E+00	0.286	-.328E+01	1.350
1.600	-.364E+00	0.266	-.327E+01	1.350
1.700	-.379E+00	0.304	-.332E+01	1.350
1.800	-.386E+00	0.293	-.326E+01	1.350
1.900	-.444E+00	0.314	-.320E+01	1.350
2.000	-.415E+00	0.285	-.315E+01	1.350
2.100	-.417E+00	0.278	-.311E+01	1.350
2.200	-.380E+00	0.249	-.306E+01	1.350
2.300	-.390E+00	0.229	-.298E+01	1.350
2.400	-.357E+00	0.232	-.302E+01	1.350
2.500	-.366E+00	0.197	-.289E+01	1.350
2.600	-.467E+00	0.195	-.275E+01	1.350

[a] Effective thickness, 0.00358 cm.

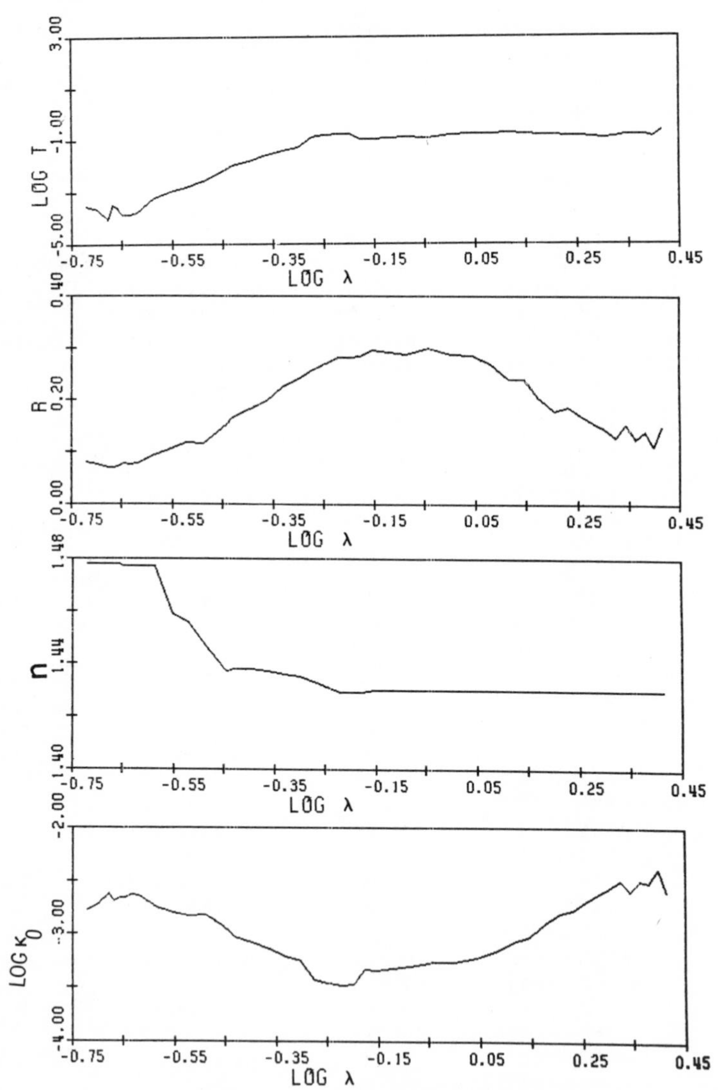

FIG. V-20 Schist, mica, French Bureau of Mines, France CP.

TABLE V-22 Schist, Mica, French Bureau of Mines, France CP[a]

LAMBDA	LOG T	R	LOG K_0	n
0.185	-.350E+01	0.079	-.278E+01	1.478
0.190	-.363E+01	0.076	-.272E+01	1.478
0.200	-.404E+01	0.069	-.262E+01	1.478
0.210	-.347E+01	0.071	-.269E+01	1.478
0.215	-.361E+01	0.071	-.266E+01	1.478
0.220	-.384E+01	0.078	-.266E+01	1.477
0.225	-.384E+01	0.076	-.263E+01	1.477
0.233	-.378E+01	0.079	-.264E+01	1.477
0.240	-.320E+01	0.095	-.275E+01	1.477
0.260	-.294E+01	0.106	-.280E+01	1.459
0.280	-.281E+01	0.118	-.283E+01	1.456
0.300	-.258E+01	0.115	-.282E+01	1.447
0.325	-.213E+01	0.151	-.297E+01	1.437
0.360	-.197E+01	0.165	-.303E+01	1.438
0.370	-.179E+01	0.181	-.308E+01	1.438
0.400	-.155E+01	0.198	-.314E+01	1.437
0.433	-.142E+01	0.225	-.321E+01	1.436
0.466	-.128E+01	0.240	-.325E+01	1.435
0.500	-.860E+00	0.258	-.343E+01	1.433
0.533	-.796E+00	0.270	-.346E+01	1.431
0.566	-.764E+00	0.281	-.348E+01	1.429
0.600	-.742E+00	0.281	-.347E+01	1.429
0.633	-.979E+00	0.283	-.333E+01	1.429
0.666	-.959E+00	0.295	-.334E+01	1.430
0.700	-.866E+00	0.287	-.330E+01	1.430
0.817	-.910E+00	0.299	-.326E+01	1.430
0.907	-.788E+00	0.288	-.326E+01	1.430
1.000	-.770E+00	0.286	-.322E+01	1.430
1.105	-.745E+00	0.269	-.316E+01	1.430
1.200	-.733E+00	0.240	-.307E+01	1.430
1.303	-.764E+00	0.239	-.302E+01	1.430
1.400	-.793E+00	0.201	-.289E+01	1.430
1.500	-.796E+00	0.179	-.280E+01	1.430
1.600	-.842E+00	0.187	-.277E+01	1.430
1.700	-.857E+00	0.169	-.269E+01	1.430
1.800	-.879E+00	0.155	-.262E+01	1.430
1.900	-.912E+00	0.145	-.256E+01	1.430
2.000	-.893E+00	0.128	-.249E+01	1.430
2.100	-.793E+00	0.152	-.259E+01	1.430
2.200	-.788E+00	0.124	-.249E+01	1.430
2.300	-.810E+00	0.139	-.251E+01	1.430
2.400	-.863E+00	0.111	-.238E+01	1.430
2.500	-.635E+00	0.148	-.260E+01	1.430
2.600	-.650E+00	0.221	-.259E+01	1.362

[a] Effective thickness, 0.00425 cm.

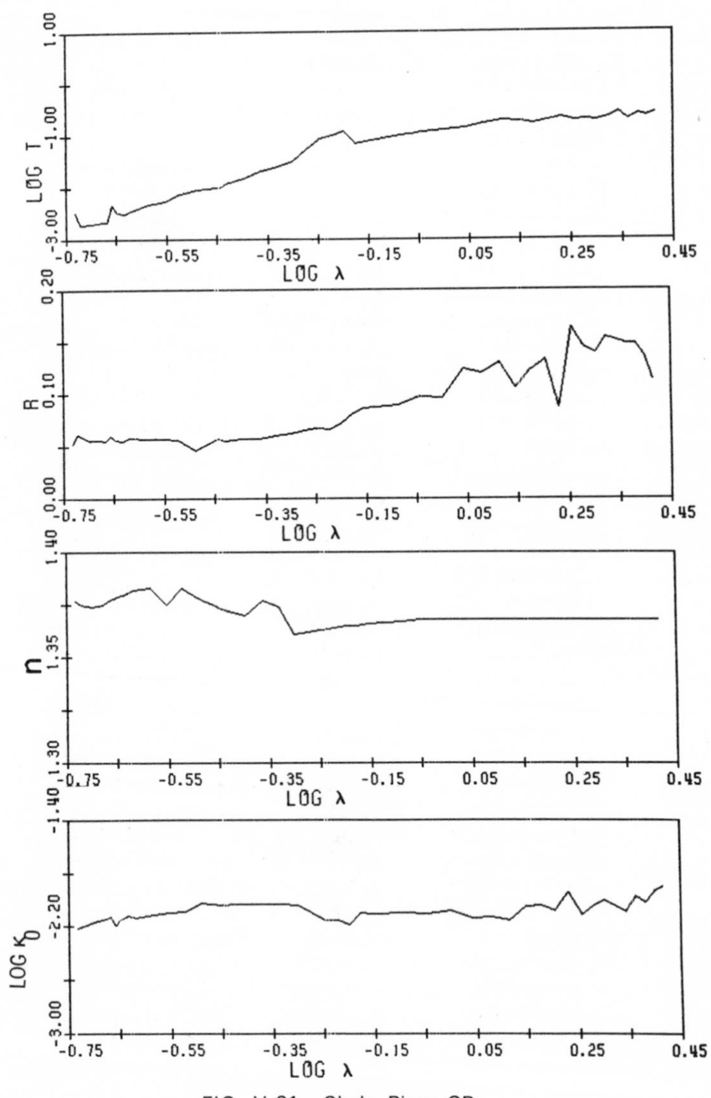

FIG. V-21 Shale, Pierre CP.

TABLE V-23 Shale, Pierre CP[a]

LAMBDA	LOG T	R	LOG K_0	n
0.185	-.246E+01	0.052	-.222E+01	1.377
0.190	-.272E+01	0.061	-.220E+01	1.375
0.200	-.269E+01	0.056	-.216E+01	1.374
0.210	-.266E+01	0.056	-.215E+01	1.375
0.215	-.266E+01	0.055	-.213E+01	1.377
0.220	-.234E+01	0.060	-.220E+01	1.378
0.225	-.247E+01	0.056	-.215E+01	1.379
0.233	-.249E+01	0.055	-.212E+01	1.380
0.240	-.243E+01	0.058	-.213E+01	1.382
0.260	-.231E+01	0.057	-.212E+01	1.383
0.280	-.226E+01	0.057	-.209E+01	1.375
0.300	-.212E+01	0.056	-.209E+01	1.383
0.325	-.205E+01	0.045	-.202E+01	1.378
0.360	-.201E+01	0.057	-.204E+01	1.373
0.370	-.192E+01	0.055	-.204E+01	1.372
0.400	-.182E+01	0.057	-.204E+01	1.370
0.433	-.167E+01	0.057	-.204E+01	1.377
0.466	-.161E+01	0.060	-.203E+01	1.374
0.500	-.150E+01	0.062	-.204E+01	1.361
0.533	-.127E+01	0.065	-.210E+01	1.362
0.566	-.107E+01	0.067	-.215E+01	1.363
0.600	-.996E+00	0.066	-.215E+01	1.364
0.633	-.921E+00	0.072	-.219E+01	1.365
0.666	-.115E+01	0.081	-.210E+01	1.365
0.700	-.113E+01	0.086	-.210E+01	1.366
0.817	-.100E+01	0.089	-.210E+01	1.367
0.907	-.936E+00	0.097	-.211E+01	1.368
1.000	-.896E+00	0.096	-.208E+01	1.368
1.105	-.866E+00	0.124	-.213E+01	1.368
1.200	-.785E+00	0.120	-.213E+01	1.368
1.303	-.724E+00	0.130	-.215E+01	1.368
1.400	-.730E+00	0.105	-.205E+01	1.368
1.500	-.783E+00	0.123	-.204E+01	1.368
1.600	-.717E+00	0.133	-.208E+01	1.368
1.700	-.670E+00	0.087	-.194E+01	1.368
1.800	-.712E+00	0.163	-.211E+01	1.368
1.900	-.710E+00	0.145	-.204E+01	1.368
2.000	-.717E+00	0.139	-.200E+01	1.368
2.100	-.668E+00	0.154	-.205E+01	1.368
2.200	-.564E+00	0.151	-.209E+01	1.368
2.300	-.699E+00	0.148	-.197E+01	1.368
2.400	-.595E+00	0.148	-.202E+01	1.368
2.500	-.644E+00	0.136	-.194E+01	1.368
2.600	-.583E+00	0.113	-.189E+01	1.368

[a] Effective thickness, 0.00101 cm.

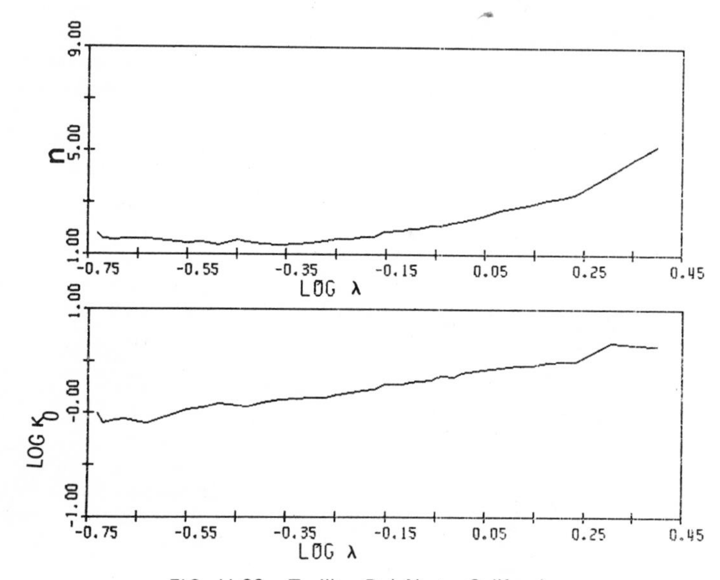

FIG. V-22 Troilite, Del Norte, California.

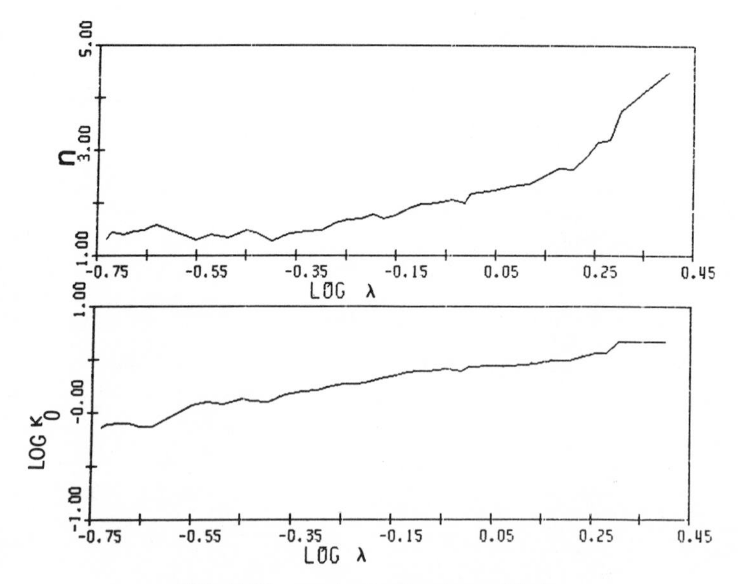

FIG. V-23 Pyrrhotite, Zacatecas, Mexico.

TABLE V-24 Troilite, Del Norte, California[a]

LAMBDA	n	K_o
0.185	1.780	0.970
0.190	1.570	0.780
0.200	1.550	0.840
0.210	1.590	0.860
0.220	1.600	0.830
0.233	1.600	0.790
0.280	1.430	1.060
0.300	1.460	1.120
0.325	1.360	1.220
0.355	1.530	1.150
0.370	1.470	1.140
0.400	1.400	1.250
0.433	1.350	1.330
0.466	1.390	1.350
0.500	1.430	1.400
0.533	1.500	1.400
0.566	1.570	1.490
0.600	1.580	1.560
0.633	1.660	1.630
0.666	1.650	1.680
0.700	1.860	1.900
0.759	1.910	1.890
0.789	1.990	1.980
0.817	1.970	2.000
0.870	2.090	2.080
0.907	2.110	2.250
0.967	2.230	2.210
1.000	2.280	2.430
1.105	2.450	2.610
1.200	2.700	2.710
1.303	2.820	2.850
1.400	2.930	2.830
1.500	3.090	3.020
1.600	3.200	3.100
1.700	3.290	3.120
2.000	4.100	4.710
2.500	5.170	4.410

[a] Data taken, in part, from Egan and Hilgeman (1975b) and Egan and Hilgeman (1977b).

TABLE V-25 Pyrrhotite, Zacatecas, Mexico[a]

n	K_o
1.320	0.720
1.430	0.770
1.390	0.790
1.450	0.790
1.480	0.740
1.570	0.730
1.300	1.170
1.400	1.250
1.330	1.190
1.480	1.330
1.440	1.280
1.280	1.240
1.420	1.470
1.460	1.560
1.470	1.610
1.610	1.760
1.680	1.840
1.700	1.860
1.770	1.940
1.700	2.080
1.750	2.140
1.920	2.350
1.970	2.410
1.980	2.430
2.010	2.470
2.050	2.520
2.000	2.450
2.180	2.670
2.240	2.740
2.320	2.750
2.360	2.810
2.520	2.940
2.650	3.070
2.630	3.050
2.880	3.360
3.760	4.620
4.500	4.620

[a] Data taken, in part, from Egan and Hilgeman (1975b) and Egan and Hilgeman (1977b).

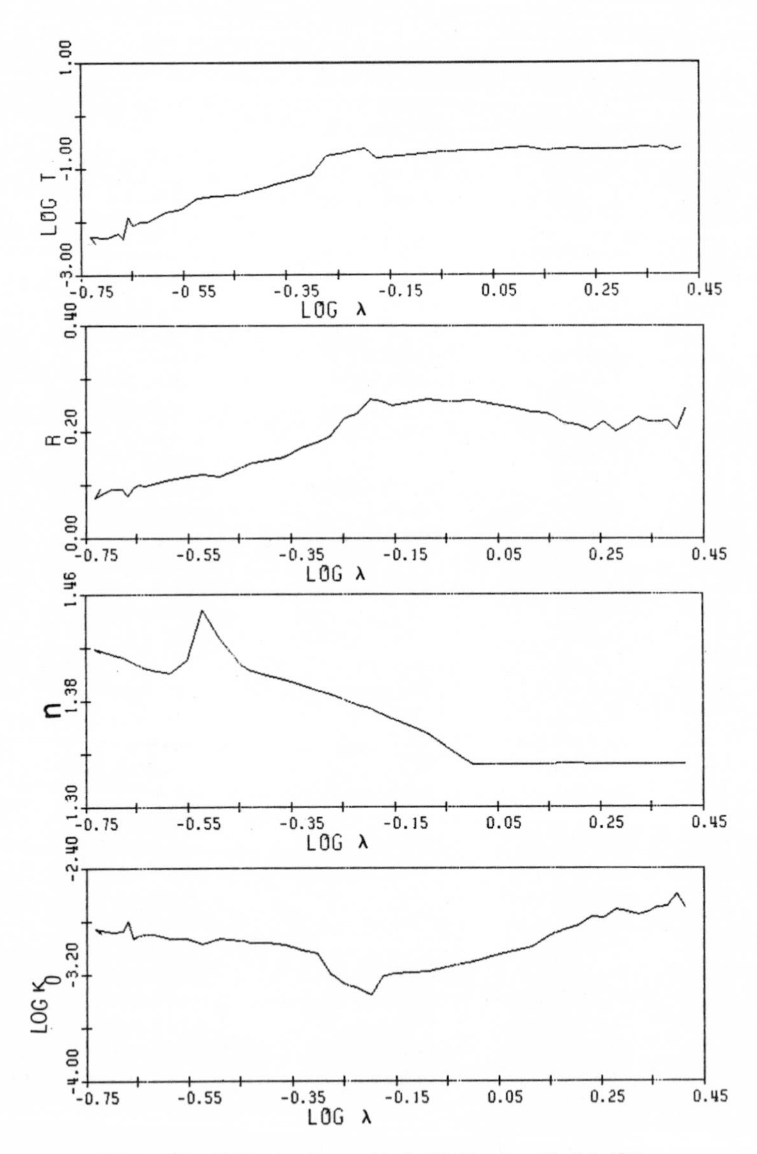

FIG. V-24 Tuff, rhyolite welded, Mt. Rogers, Virginia CP.

TABLE V-26 Tuff, Rhyolite Welded, Mt. Rogers, Virginia CP[a]

LAMBDA	LOG T	R	LOG K_0	n
0.185	-.241E+01	0.094	-.289E+01	1.417
0.185	-.229E+01	0.076	-.285E+01	1.419
0.200	-.230E+01	0.091	-.288E+01	1.415
0.210	-.223E+01	0.091	-.287E+01	1.413
0.215	-.233E+01	0.079	-.279E+01	1.411
0.220	-.190E+01	0.093	-.292E+01	1.409
0.225	-.206E+01	0.099	-.290E+01	1.407
0.233	-.200E+01	0.099	-.290E+01	1.405
0.240	-.200E+01	0.102	-.290E+01	1.403
0.260	-.183E+01	0.111	-.293E+01	1.401
0.280	-.175E+01	0.115	-.293E+01	1.410
0.300	-.155E+01	0.120	-.297E+01	1.448
0.325	-.151E+01	0.115	-.293E+01	1.427
0.360	-.150E+01	0.134	-.294E+01	1.406
0.370	-.145E+01	0.140	-.296E+01	1.403
0.400	-.138E+01	0.146	-.296E+01	1.399
0.433	-.127E+01	0.152	-.298E+01	1.396
0.466	-.120E+01	0.171	-.302E+01	1.392
0.500	-.112E+01	0.180	-.304E+01	1.388
0.533	-.767E+00	0.195	-.320E+01	1.385
0.566	-.717E+00	0.225	-.327E+01	1.381
0.600	-.664E+00	0.236	-.330E+01	1.377
0.633	-.629E+00	0.261	-.336E+01	1.374
0.666	-.801E+00	0.258	-.322E+01	1.370
0.700	-.777E+00	0.249	-.319E+01	1.366
0.817	-.714E+00	0.261	-.318E+01	1.355
0.907	-.686E+00	0.257	-.314E+01	1.343
1.000	-.664E+00	0.259	-.311E+01	1.332
1.105	-.658E+00	0.252	-.305E+01	1.332
1.200	-.623E+00	0.246	-.303E+01	1.332
1.303	-.595E+00	0.238	-.300E+01	1.332
1.400	-.662E+00	0.237	-.292E+01	1.332
1.500	-.631E+00	0.217	-.286E+01	1.333
1.600	-.622E+00	0.214	-.283E+01	1.333
1.700	-.640E+00	0.203	-.277E+01	1.332
1.800	-.648E+00	0.219	-.278E+01	1.332
1.900	-.648E+00	0.201	-.271E+01	1.332
2.000	-.633E+00	0.213	-.273E+01	1.332
2.100	-.616E+00	0.228	-.275E+01	1.332
2.200	-.590E+00	0.220	-.273E+01	1.332
2.300	-.618E+00	0.221	-.269E+01	1.332
2.400	-.607E+00	0.222	-.269E+01	1.332
2.500	-.660E+00	0.205	-.259E+01	1.332
2.600	-.614E+00	0.245	-.270E+01	1.332

[a] Effective thickness, 0.00326 cm.

CHAPTER VI □ SURFACE REFLECTANCE MODELING USING COMPLEX INDICES

A. INTRODUCTION

The previous chapters have described techniques for the determination of the optical complex indices of nonhomogeneous materials. A representative listing of the indices was also presented. Once these fundamental optical properties are known, they may be used in an appropriate scattering model, such as one of those described in Chapter II, to predict surface reflectance. This chapter will utilize a variety of approaches to model surfaces starting with the Mie theory, then two- and six-flux models, and, finally, a noncontinuum model. We have made use of available computer models, when possible, and the present objective is the description of surface reflection in the wavelength range from the ultraviolet to the near infrared. Various scales of surface modeling will be described, from macrosurfaces, seen by satellite or aircraft to microsurfaces microns in size, observed with a laboratory photometer.

B. MIE THEORY

The Mie theory (presented in Chapter II) describes the scattering and absorption of radiation incident on single spherical particles. If the spheres make up a hazy atmosphere or a more dense surface, some modification must be added to account for the radiative transfer between particles. Also, naturally occurring particles are not actually spheres, and generally have a range of sizes. The nonsphericity of the particles (i.e., the presence of edges and asperities on the particles) as well as the presence of a particle size distribution may be taken approximately into account together by using a particle size distribution appropriately weighted by particle size. In other words, small particles can produce a scattering resembling edges and asperities, and the effect can be approximated by a high concentration of very small particles.

For simplicity, an illustration of surface modeling will be indicated in this section that is limited to one characteristic particle size (6-μm diameter spheres). We shall calculate the upward flux for two models consisting of nonhomogeneous material. The two models are

(1) the upward flux from a "cloudy" atmosphere above a Lambertian scattering surface; and

(2) the radiation scattered from a dense (6-μm diameter) particulate surface.

The nonhomogeneous material used is the Bruderheim meteorite, for which optical complex indices of refraction have been presented in Chapter V.

In general, the use of the Mie theory to represent the reflectance and absorption (or emissivity) of surfaces in the infrared is not new. Conel (1969) describes the use of the Mie theory to develop a cloudy atmosphere model for radiative transfer in a condensed powder; his concern is the wavelength range between 7 and 17 μm; however, his approach may be extended from the near infrared to the ultraviolet with an appropriate cloudy atmosphere model. A readily available Mie scattering program that conveniently allows the manipulation of the many scattering parameters was developed by Dave (1972) of IBM. His four technical reports describe the programs developed under NASA sponsorship (Contract No. NAS5-21680). The overall program, in FORTRAN, prints the flux and intensity of scattered radiation emerging from a plane-parallel nonhomogeneous atmosphere at any level. If desired, an arbitrary vertical distribution and concentration of Mie scattering particulates may be used. Provision is also made for adding the effects of molecular absorption and scattering. The computation method employs a representation of the scattering phase matrix obtained from a variation of a modified Fourier series (Dave, 1970a).

The first report (Program I), describes the Legendre series basic coefficients representing a scattering phase function for spherical particles. This program is used to obtain the radiation scattered by a single sphere through scattering angles from 0 to 180°.

The second report (Program II) describes the calculation of the Legendre series coefficients of a unit-volume normalized scattering phase function from the single-sphere functions. The volume is assumed illuminated by unpolarized, unidirectional monochromatic radiation. The particle size distribution is arbitrary, but the index of refraction must be the same for all particles. For nonhomogeneous materials, an average index of refraction must be used. The output is used as input to Program III.

The third report (Program III) describes the transformation of the coordinates of the scattering phase function in terms of incident and emergent zenith angles. The Fourier coefficients of the normalized scattering phase function series of a unit volume are computed. The coefficients are computed in 2° angular steps. The output is then used as input to Program IV.

The fourth report (Program IV) describes the utilization of the basic radiative transfer equation with a layered atmosphere model. The radiative transfer equation is decomposed into a series of independent integrodifferential equations representing the scattered intensity transfer by the nth Fourier component. The Fourier series argument is the difference between the azimuthal angles of the planes containing the incident and emergent rays. Each integrodifferential equation is individually solved using the Gauss–Seidel iterative procedure for an atmosphere divided into a finite number of layers.

The use of this program for a particle diameter of 6 μm will now be described. The results of the computations will be compared to spectrogoniometric surface measurements using a large-scale photometer polarimeter (i.e., one with a viewing area of $\sim$70-mm diameter).

1. Cloudy Atmosphere Model

The calculations have been made between 0.280- and 1.105-μm wavelengths, using the optical complex index of refraction of the low iron–nickel sample of the Bruderheim meteorite (compressed pellet sample of Fig. V-15 and Table V-17). The 6-μm particle size was assumed as an approximate representation of the <37-μm particles (Fig. VI-1).

A cloudy atmosphere model may now be produced by the Dave program (Program IV); Program II is bypassed (because the particle size distribution is not utilized), and Program III produces the Fourier coefficients required for Program IV. This part of the Dave program was developed for atmospheric applications with distances scaled in kilometers above a reference plane. The phenomena scale as long as the phase effects in particle-to-particle interactions are ignored. For convenience, we have kept the Dave convention for the calculations. The vertical particle density distribution for a 40-layer scattering model is shown in Fig. VI-2. This "cloudy atmosphere" is taken to roughly represent a surface, with the calculated upward flux results presented in Fig. VI-1 (as squares). The theoretical curve could be raised by using smaller particle sizes, but the general shape would not significantly change. A considerable amount of computing time is involved in the Mie calculations.

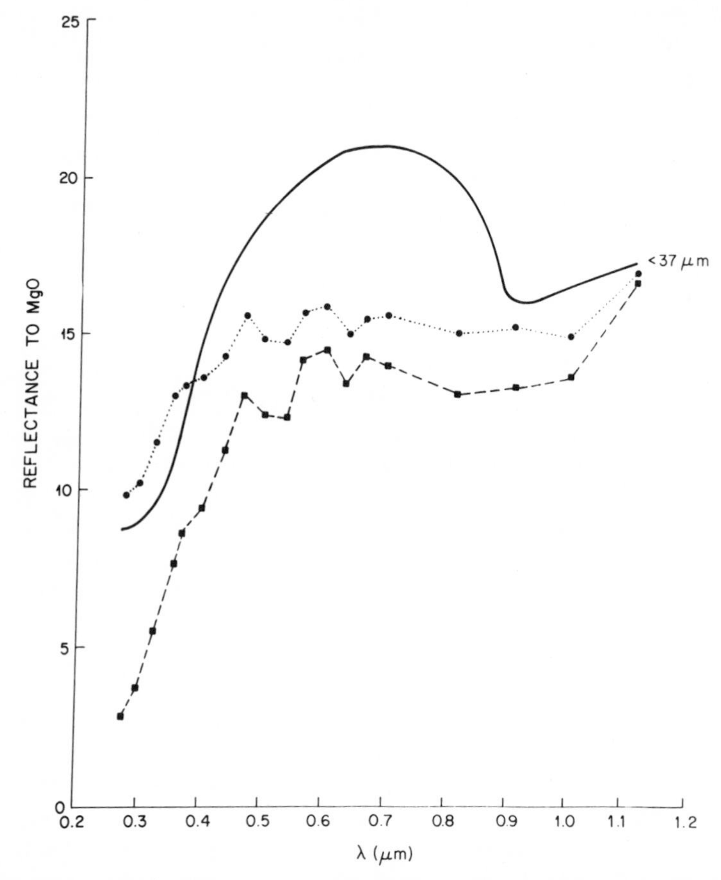

FIG. VI-1 Reflectance as a function of wavelength observed for < 37-μm sized Bruderheim meteorite particles (———), compared to calculated values: ■, upward flux from a cloudy atmosphere above a Lambertian surface; ●, radiation scattered from a dense 6-μm particulate surface.

2. Dense Particle Surface Model

Now let us consider the effect of a drastic decrease in the particle density above the 19th layer, with no particles above, to produce an actual surface (Fig. VI-2, dotted line). The calculated upward fluxes are also shown in Fig. VI-1 (as solid circles). The agreement is slightly better, but still leaves a lot to be desired. More detailed models are possible; combinations of particle sizes and distributions could help in improving the match.

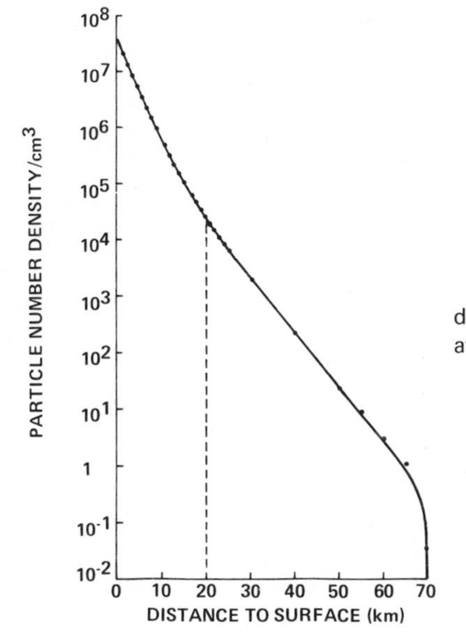

FIG. VI-2 Vertical particle density distribution used to produce a cloudy atmosphere model.

C. TWO-FLUX MODEL

The two-flux model (see Chapter II) has been widely adopted for the analytical determination of absorption and scattering because of its simplicity. However, in using it for the inverse problem, surface modeling, the application is limited. Two versions are commonly used for absorption and scattering determinations: The Kubelka–Munk theory (KM) (Kubelka and Munk, 1931; Kubelka, 1948) and the modified Kubelka–Munk (MKM) theory (Reichman, 1973; Egan *et al.*, 1973a). Both theories require the absorption and scattering coefficients to be specified for a surface, and the MKM theory further requires the refractive portion of the optical complex index of refraction. The geometry of the particles in the surface is immaterial to the calculation, being implicit in the specification of the scattering coefficient. The application of either theory to surface modeling requires an assumption about the scattering characteristics of a surface. The density of the scatterers (and hence the absorption) must be accounted for with an apparent absorption coefficient for the surface. It should be noted that the KM theory fails experimentally for large absorption coefficients because of the neglect of surface reflection (Vincent and Hunt, 1968). The MKM theory corrects for this by taking surface reflection into account. An example of the application of the MKM theory was presented in Chapter II, so no example will be given here.

D. SIX-FLUX MODELS

A number of multiflux models have been proposed (Schuster, 1905; Kubelka and Munk, 1931; Chandrasekhar, 1960; Plass and Kattawar, 1968, 1971; Blevin and Brown, 1961; Emslie and Aronson, 1973; Egan and Hilgeman, 1978a), of these two (six-flux models) have been found to be very useful for surface modeling: one is that of Emslie and Aronson (1973) and the other is that of Egan and Hilgeman (1978a). Both theories require the specification of a particle nominal diameter (although weighted distributions of particle sizes may be used) and a specification of the absorption by edges and asperities. The Monte Carlo theory of Egan and Hilgeman (1978a) further requires the specification of the ratio of internal to external scattering and the scattering by edges and asperities.

The two theories may be compared to model the Bruderheim meteorite in the wavelength range 0.28–1.105 μm. A comparison of the calculations using the two theories is presented in Fig. VI-3 for <37-μm Bruderheim meteorite particles. A good match with the Emslie and Aronson theory calculations (shown as solid circles) could not be obtained. Also a curve tilt upward occurs with increase in wavelength. The tilt appears to be due to omission of scattering by asperities; their theory was developed to explain the infrared reflectance of surfaces, where scattering by edges and asperities is minimal. However, when the theory is applied in the ultraviolet and visual optical ranges, the neglect of scattering by edges and asperities can cause the observed tilt in the calculated results.

The Monte Carlo model of Egan and Hilgeman (1978a) produces a better match (shown as crosses in Fig. VI-3). Figure VI-3 is a comparison of the experimental curves for the <37-μm particle sizes with the Monte Carlo calculations for a 3-μm average particle size with low external scattering ($SI = 0.70$). The fit is best between 0.28 and 0.600 μm, with a poorer fit at longer wavelengths; peaking is lacking at 0.700 μm, and the dip at 0.907 μm is not quite enough.

Further application of the Monte Carlo model permits closely matching the measured reflectance of 75–250 μm, uncoated 0.25 to 4.76-mm particles, and 0.25–4.76-mm particles coated with <74-μm particles of the Bruderheim meteorite (Figs. VI-4–VI-6). The calculations for the <37-μm and the 0.25–4.76-μm coated-particle curves were made with the low iron–nickel indices of refraction (compressed pellet of Fig. V-15 and Table V-17) because of the lack of iron–nickel in the finer powders. The calculations for the 75–250-μm and the 0.25–4.76-mm uncoated particles were made with the high iron–nickel indices of refraction (thin section of Fig. V-16 and Table V-18).

Referring to Fig. VI-4, which was calculated for 70-μm particles using the high iron–nickel indices of refraction (thin section sample of Fig. V-16 and

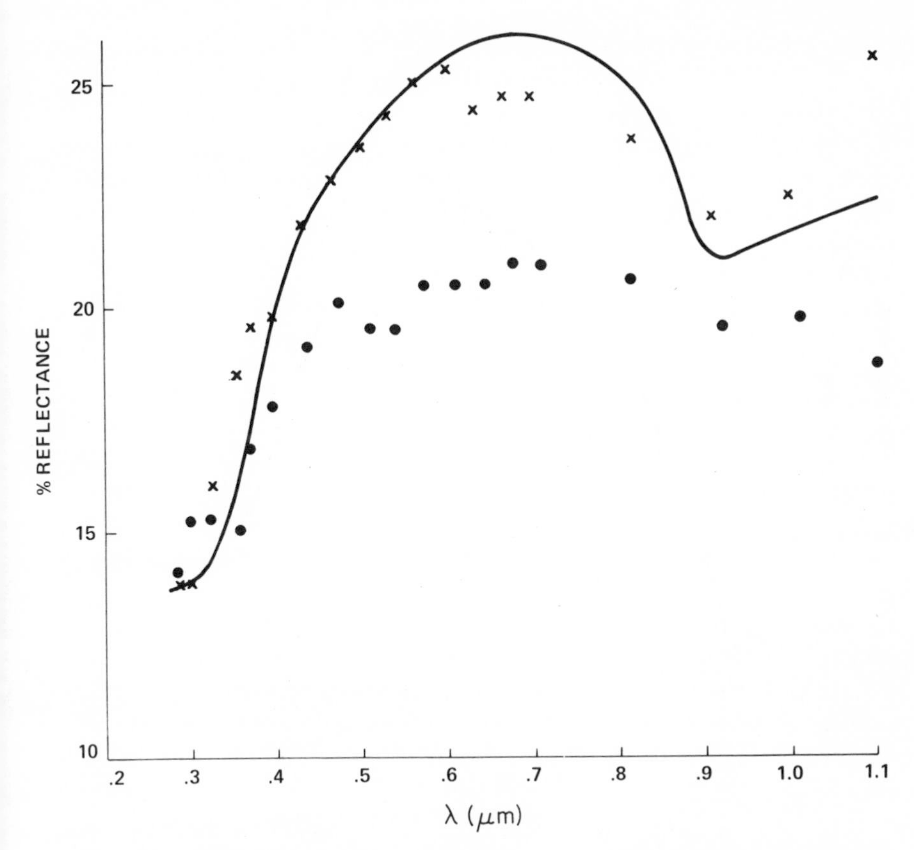

FIG. VI-3 Comparison of the experimental (——), theoretical Monte Carlo (×), and Emslie–Aronson (●) reflection spectra for <37-μm sized Bruderheim meteorite particles between 0.28- and 1.105-μm wavelengths; $d = 3$ μm, $\hat{x} = \kappa_0/\lambda$, $\hat{y} \propto \lambda^{-1/3}$, SI = 0.70. [From Egan and Hilgeman (1978a).]

Table V-18), the fit is much better at wavelengths longer than 0.600 μm. The <37-μm low iron–nickel index fit could be improved by adding a contribution from the high iron–nickel indices of refraction, but the uniqueness of the fit is then open to serious objections. The possibility of combining indices of refraction, as well as various particle size curves, would permit fitting the experimental curves with many alternative selections. However, what was sought was the simplest match in terms of a reasonable particle size and appropriate indices of refraction.

The theoretical reflection spectrum of the 0.25–4.76-mm particles (Fig. VI-5), calculated using the high index of refraction, is a bit too peaked and could likewise benefit from a contribution of the indices of refraction

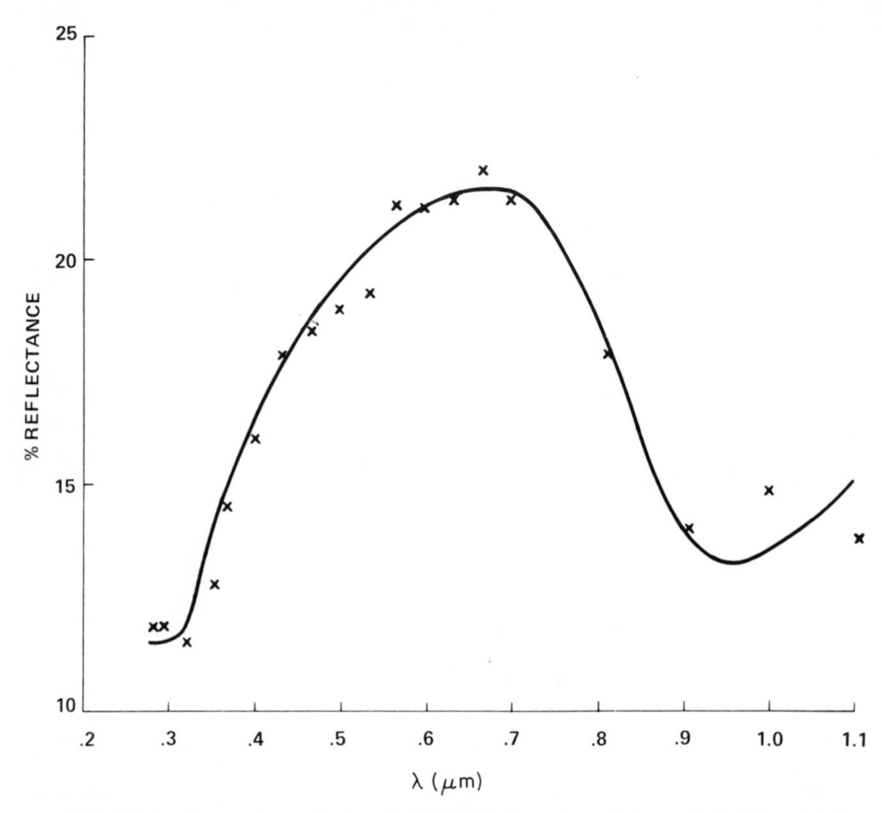

FIG. VI-4 Comparison of the experimental (——) and theoretical (×) reflection spectra for 75–250-μm sized Bruderheim meteorite particles between 0.28- and 1.105-μm wavelengths; $d = 70$ μm, $\hat{x} = \kappa_0/\lambda$, $\hat{y} \propto \lambda^{-1/3}$, SI = 0.37. [From Egan and Hilgeman (1978a).]

of the low iron–nickel sample; the flattening is probably the result of some low iron–nickel fines adhering to the particle surfaces.

The reflection spectrum of the 0.25–4.76-mm particles dusted with <74-μm particles is fairly well fitted with the low iron–nickel indices of refraction (Fig. VI-6); the smaller, brighter low iron–nickel fines dominate the reflectance.

The four particulate surfaces in this example have nearly Lambertian scattering properties, with some small geometrical differences in scattering that are dependent upon wavelength and the phase angle (the angle between the incident and scattered rays). To a first approximation, we assume the surfaces to be Lambertian, and thus an equivalence can be made between the bidirectional reflectance and the six-flux model calculated reflectance.

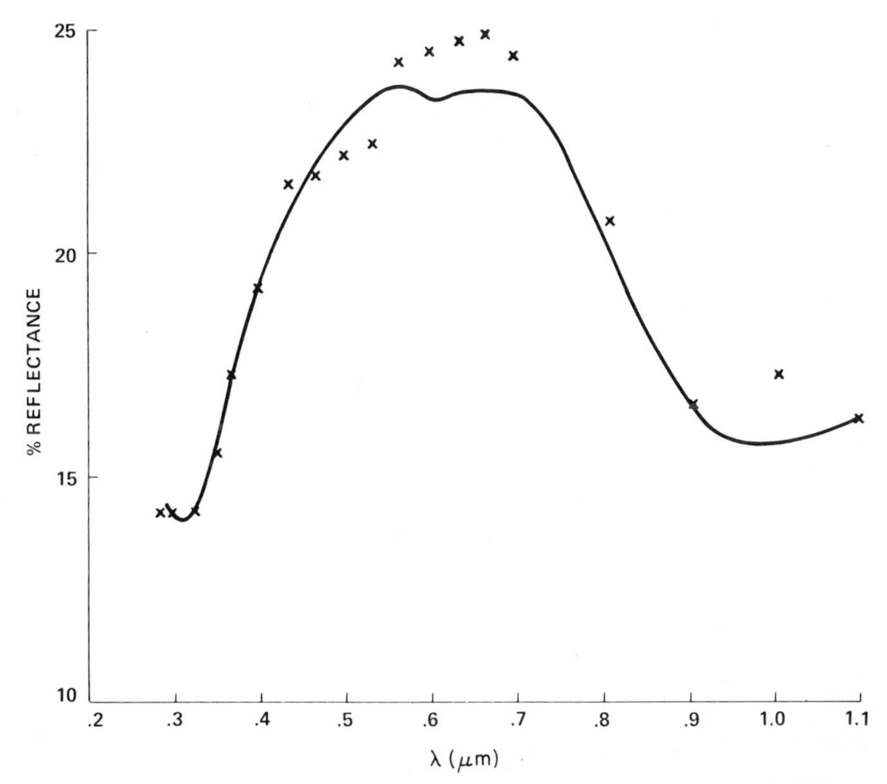

FIG. VI-5 Comparison of the experimental (——) and theoretical (×) reflection spectra for 0.25–4.76-mm sized Bruderheim meteorite particles between 0.28- and 1.105-μm wavelengths; $d = 50$ μm, $\hat{x} = \kappa_0/\lambda$, $\hat{y} \propto \lambda^{-1/3}$, SI = 0.30. [From Egan and Hilgeman (1978a).]

The absolute reflectance of the four size-range samples of the Bruderheim meteorite was measured previously in the range of wavelengths from 0.3 to 1.1 μm for normal incidence, 3.6° scattering angle.

The Monte Carlo theory is restricted to particles of size (diameter) larger than the wavelength (where a geometrical optics approach is valid). For particles smaller than the wavelength (such as dust), the Emslie–Aronson small-particle theory must be used.

Single average particle size fits were used to increase the uniqueness, and, as a secondary benefit, to hold down the computer time. The fit is good for reasonable particle sizes in Figs. VI-3, VI-4, and VI-6; however, in Fig. VI-5 this is not the case. The reason is evident from microscopic observations of the 0.25–4.76-mm particles, which reveal the existence of asperities and small-scale structure.

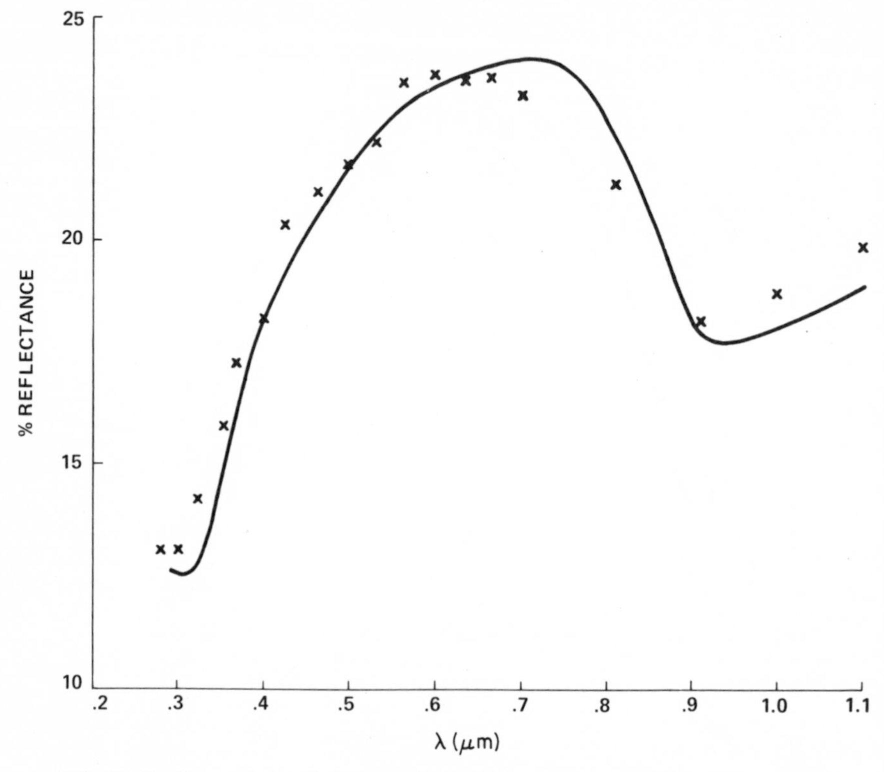

FIG. VI-6 Comparison of the experimental (——) and theoretical (×) reflection spectra for 0.25–4.76-mm sized Bruderheim meteorite particles, coated with <74-μm particles, between 0.28- and 1.105-μm wavelengths; $d = 3$ μm, $\hat{x} = \kappa_0/\lambda$, $\hat{y} \propto \lambda^{-1/3}$, SI = 0.75. [From Egan and Hilgeman (1978a).]

E. NONCONTINUUM MODELS

A noncontinuum model treats the scattering and absorption as interrupted spatially, even when the powders are in contact. The approach described by Egan and Hilgeman (1978a) in the six-flux model is a noncontinuum model as far as the scattering is concerned for one particle; however, the individual-particle scattering is combined into a continuum theory to treat the radiative transfer problem. Melamed (1963) continues a similar separate-particle approach into a summation of the multiply reflected radiation between the particles to get the diffuse reflectance. However, the complex nature of the optical index of refraction is taken into account only in the transmission of radiation through a particle producing absorption and in reducing the strength of the emergent refracted ray. Also, the refraction of the emerging

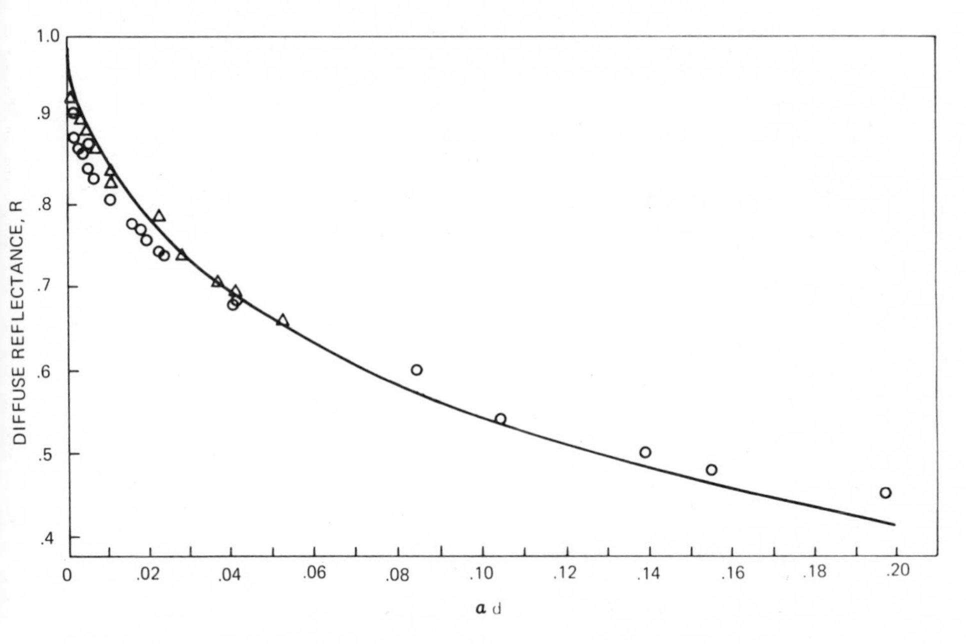

FIG. VI-7 Dependence of reflectance R on αd for didymium glass; △, measured values, 34-μm diameter particles; ○, measured values, 128-μm diameter particles; ——, theoretical curve for $n = 1.55$. [From Melamed (1961).]

ray is neglected. The effect of particle surface irregularities is grouped under hypothetical reflection coefficients for internal and external rays. Good agreement for large particle size is expected and is seen in the experimental variation of diffuse reflectance with absorption coefficient α using a specific material, didymium glass (Fig. VI-7). The quantity d is the particle diameter, approximated by a sphere of volume equal to that of the average particle. The relation to the physical scattering and absorption phenomena does not appear in this approach.

The approach used by Lathrop (1966) assumes parallel sheets of homogeneous layers of thin finite thickness. The requirement is the prediction of the absorption as a function of the mass per unit area. The effects of particle geometry and shape do not explicitly appear in the theory and neither does their interrelationship. Thus, the specific effects of variations in particle size, shape, spacing, absorption, or surface roughness cannot be seen.

NOTES ON SUPPLEMENTARY READING MATERIAL

Deirmendjian (1969): An introduction to single and many particle scattering; applications to terrestrial and planetary atmospheric problems.

Mason (1962): Presents a comprehensive account of meteorites, with special emphasis on their chemical and mineralogical composition, and their structure.

McLaughlin (1961): A good introduction to astronomy; chapter on physical features and atmospheres of planets.

Wendlandt and Hecht (1966): Presents the two-flux model of a scattering medium; many applications discussed along with instrumentation.

CHAPTER VII □ SURFACE RETROREFLECTANCE (OPPOSITION EFFECT)

A. INTRODUCTION

Theories of diffuse surface reflectance have been presented in Chapter VI; however, even with very diffuse surfaces, deviations from diffuseness have been observed in the retroreflection direction. This deviation from diffuseness consists of a sharp nonlinear upsurge in brightness near the retroreflection direction and has been termed the opposition effect. In particular, since diffuse surfaces are used as photometric standards and coatings, any variation from diffuseness is of considerable interest both theoretically and practically.

In order to quantify the opposition effect, various investigators have defined several opposition effect ratios such as the ratio of the retroreflection to the reflection at a larger angle or the ratio of the retroreflection to the slope of the photometric curve projected to 0° angle (Oetking, 1966; Barabashev, 1922; Markov, 1924; van Diggelen, 1965; Gehrels *et al.*, 1964; Hapke, 1963).

The glory appearing around the shadow of an airplane on a cloud, when viewed in the retroreflecting direction (i.e., from the airplane), is an example of the peak in reflectivity occurring from the opposition effect. Similar halos have been observed, varying in intensity and size, above forests, grass, or cornfields (Oetking, 1966).

In astronomy, a sharp rise in the moon's reflective intensity at full moon was observed by Barabashev (1922) and by Markov (1924). More recently, van Diggelen (1965) and Gehrels *et al.* (1964) deduced (by extrapolation) that the moon's surface brightness may increase by as much as a factor of 2 at opposition.

This chapter will review laboratory measurements of the opposition effect, first, at phase angles to within 1° of retroreflectance, and, second, at exact retroreflectance. In some of the more recent work, the effect of illuminating source coherence is investigated, through the use of lasers. Finally, various theoretical interpretations will be presented which attempt to explain the retroreflectance effect, with varying degrees of success.

B. EXPERIMENT

1. 1° Phase Angle

Oetking (1966) was the first to show the opposition effect peaking in laboratory measurements. He accomplished this by devising an optical system that permitted photometric observations to be made to within 1° of the incident direction. This was not considered to be a limitation in terms of astronomical interpretations because the minimum observable collimation angle of the solar radiation on the moon's surface as seen from the earth is limited to about 0.5° by the diameter of the sun. Oetking (1966) showed peaking for smoked MgO and $MgCO_3$ (photometric standards), Al_2O_3, and various other substances. In fact, all the material measured showed an opposition effect (Fig. VII-1). It was pointed out that the opposition effect could not be a specular reflectance effect because it occurred even when the surface was tilted at angles up to 50° from the incident direction.

2. 0° Phase Angle

Subsequently, Egan and Hilgeman (1976) used a beam-splitting technique to make retroreflectance measurements at exactly 0° incidence and viewing; they used an unorthodox location for the chopper to eliminate scattered background radiation. The surfaces measured were diffuse photometric standards and coatings such as $MgCO_3$, $BaSO_4$ paint, sulfur, Nextel white, red, blue, and black paints. They corroborated the observations of Oetking (1966), and included laser observations of the retroreflectance effect.

In the initial work of Egan and Hilgeman (1976), the source collimation and sensor angles were 1°; the retroreflectance at 0° was referenced to 30°, and is listed in Table VII-1 as a function of wavelength for incandescent illumination for $MgCO_3$, $BaSO_4$, and sulfur. The use of 30° as the reference angle is immaterial with diffuse (cosine) scattering surfaces; that is, any angle sufficiently far away from the retroreflecting direction is suitable for the reference. There is very little wavelength dependence of the opposition effect, with $BaSO_4$ having the highest.

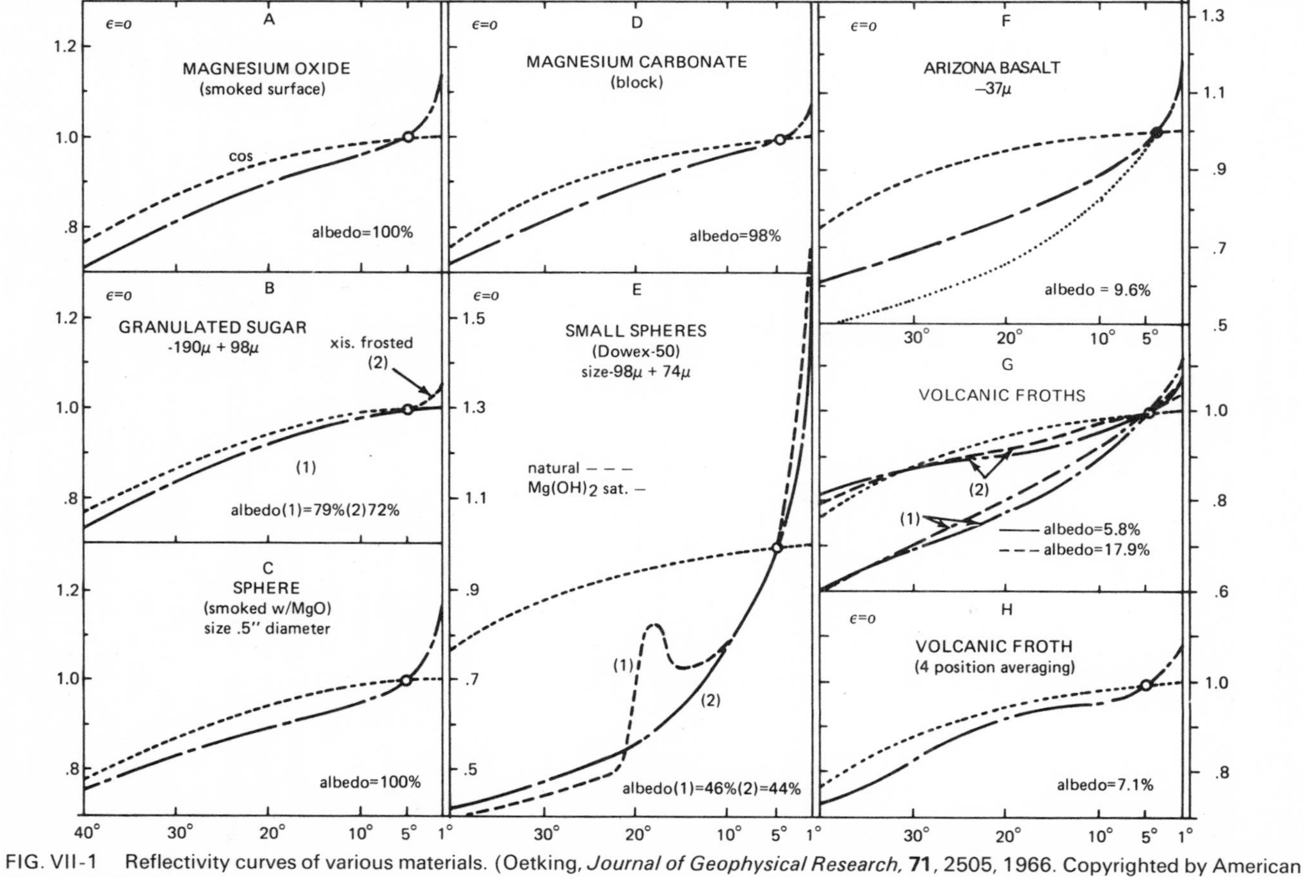

FIG. VII-1 Reflectivity curves of various materials. (Oetking, *Journal of Geophysical Research,* **71**, 2505, 1966. Copyrighted by American Geophysical Union.)

TABLE VII-1 Opposition Effect[a] Referenced to 30° for Photometric Standards[b]

λ (μm)	$MgCO_3$	$BaSO_4$	Sulfur (colloidal)
0.35	1.27	1.48	—
0.400	1.26	1.51	—
0.500	1.28	1.53	1.25
0.600	1.29	1.53	1.35
0.817	1.25	1.58	1.33
1.0	1.20	1.48	1.28

[a] 1° source and sensor collimation.
[b] From Egan and Hilgeman (1976).

TABLE VII-2 Opposition Effect[a] Referenced to 30° for 3M Nextel Paints[b]

	Paint			
λ (μm)	White	Blue	Red	Black
0.400	1.84	1.61	1.50	1.58
0.466	1.70	1.52	1.49	1.63
0.600	1.59	1.32	1.60	1.80
0.817	1.77	1.57	1.71	1.23
1.105	1.59	1.31	1.51	—

[a] 1° source and sensor collimation.
[b] From Egan and Hilgeman (1976).

Similarly, the opposition effect for the 3M Nextel paints is listed in Table VII-2; all of these paints have an opposition effect of approximately 1.5.

Now compare the retroreflectance effect using highly coherent incident illumination from a He–Ne laser at 0.6328-μm wavelength. The scattering properties of a good diffusing $BaSO_4$ surface with incandescent illumination are represented in Fig. VII-2 on a polar diagram. The polar diagram plots the brightness as a function of the scattering angle (measured from the normal to the sample) for incident radiation from a source directly above the sample. The origin of the coordinate system is taken at the sample. The retroreflectance peak is shown as well as what the scattered radiation would be from a perfect cosine diffuser. However, with the laser illumination, the scattering is depressed below the cosine curve at scattering angles below about 40° (Fig. VII-3). This effect had been observed previously by Egan

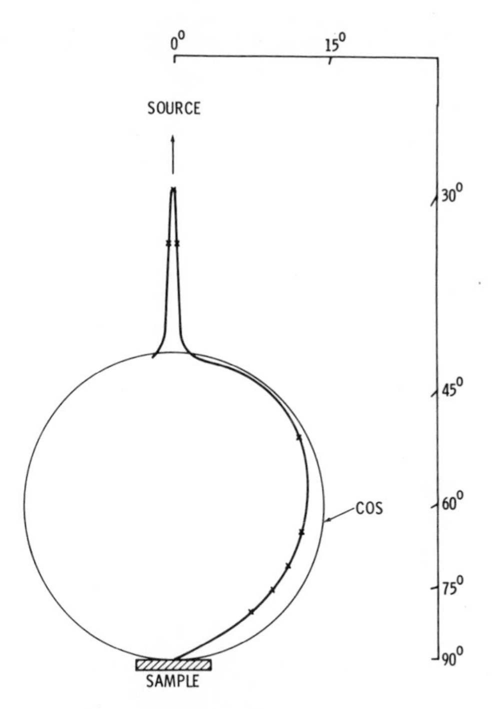

FIG. VII-2 Reflectance properties of $BaSO_4$ coating at $\lambda = 0.6$ μm with low-coherence illumination. [From Egan and Hilgeman (1976).]

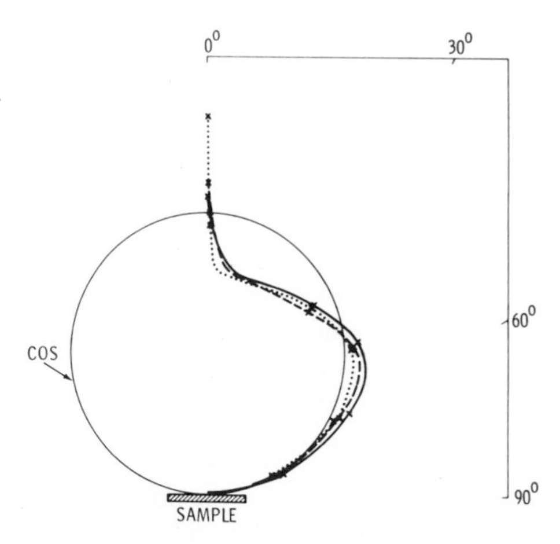

FIG. VII-3 Reflectance properties of $BaSO_4$ (–––), $MgCO_3$ (——), and sulfur (· · ·) with high-coherence illumination (0.6328-μm He–Ne laser). [From Egan and Hilgeman (1976).]

and Hallock (1969) with a large-scale photometer and a He–Ne laser. For large scattering angles, the perfect diffuser is approached. The sulfur is seen to have the highest opposition effect of the three samples. The laser retroreflectances for $MgCO_3$, $BaSO_4$, sulfur, and the Nextel paints are tabulated in Table VII-3; the values are close to those presented in Tables VII-1 and VII-2 except for the blue and black paints, which show a remarkably large opposition effect.

An investigation of the retroreflectance effect of $BaSO_4$ and $MgCO_3$ for a range of surface tilts reveals no variation with incandescent illumination with source collimation and sensor acceptance angles of 1° (Fig. VII-4); the curves have been normalized by division by the cosine of the scattering angle. Similarly, using laser illumination, these surfaces as well as sulfur and white Nextel paint reveal no variation in retroreflectance effect with scattering angle (Fig. VII-5). However, when the source collimation–sensor acceptance angle is decreased below 1°, although the exact amount is undefined, the opposition and retroreflectance effect increases for all samples (Fig. VII-6). The increase is particularly large for sulfur.

TABLE VII-3 Laser (0.6328 μm) Opposition Effect[a] Referenced to 30° for Photometric Samples[b]

Sample	Opposition effect	Sample	Opposition effect
$MgCO_3$	1.35	Nextel blue	8
$BaSO_4$	1.49	Nextel red	1.40
Sulfur	1.76	Nextel black	11
Nextel white	1.31		

[a] 1° source and sensor collimation.
[b] From Egan and Hilgeman (1976).

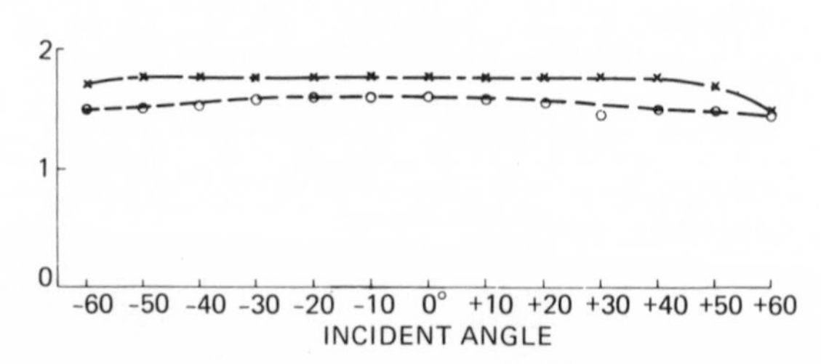

Fig. VII-4 Retroreflectance effect at varying incident angles for $BaSO_4$ (×) and $MgCO_3$ (○) for $\lambda = 0.600$-μm low-coherence illumination using 1° source collimation and sensor acceptance angles. [From Egan and Hilgeman (1977a).]

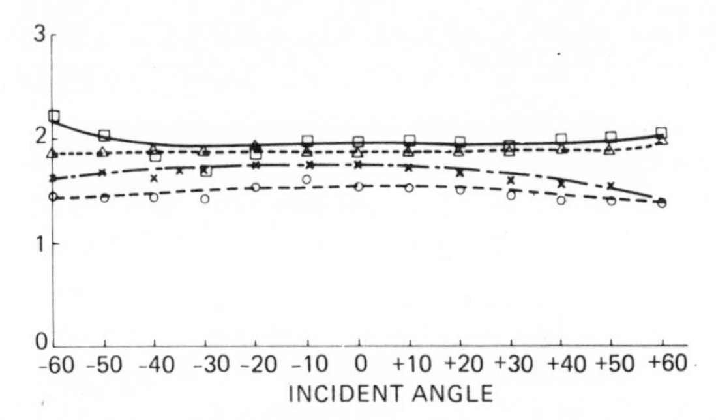

FIG. VII-5 Retroreflectance effect at varying incident angles for $BaSO_4$ (×), $MgCO_3$ (○), white Nextel paint (△), and colloidal sulfur (□) for $\lambda = 0.6328$-μm He–Ne laser illumination with 1° source collimation and sensor acceptance angles. [From Egan and Hilgeman (1977a).]

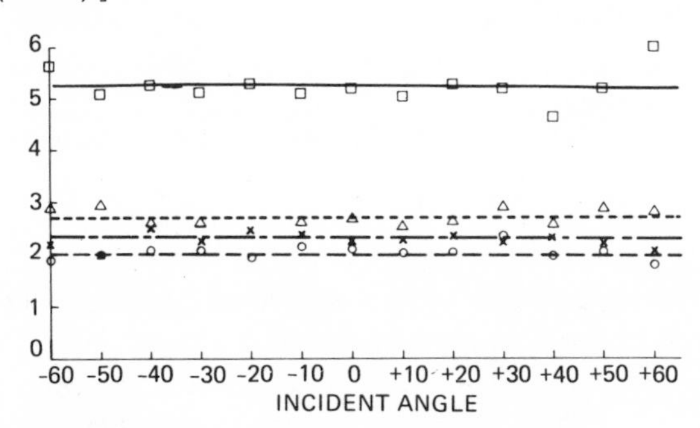

FIG. VII-6 Retroreflectance effect at varying incident angles for $BaSO_4$ (×), $MgCO_3$ (○), white Nextel paint (△), and colloidal sulfur (□) for $\lambda = 0.6328$-μm He–Ne laser illumination with better than 1° source collimation and sensor acceptance angles. [From Egan and Hilgeman (1977a).]

C. THEORY

Several theories have been proposed to represent the retroreflectance effect. These can be described as a shadowing-type theory, Mie scattering theories, or as a corner reflector mirror theory.

1. Shadowing Theory

In order to analyze the causes of the observed retroreflectance scattering properties, a shadowing theory was developed by Hapke (1963). The opposition effect was described by a function

$$B(\alpha, g) = 2 - [1/(2a)](1 - e^{-a})(3 - e^{-a}) \qquad \text{for} \quad |\alpha| \leq \pi/2 \quad \text{(VII-1)}$$

$$B(\alpha, g) = 1 \qquad \text{for} \quad |\alpha| > \pi/2 \quad \text{(VII-2)}$$

TABLE VII-4 Theoretical Opposition Effect[a]

f	Opposition effect[b]	f	Opposition effect[b]
0.01	1.98	0.30	1.55
0.05	1.90	0.40	1.47
0.10	1.80	0.50	1.29
0.20	1.65	0.60	1.18

[a] From Egan and Hilgeman, 1976.
[b] Referenced to 30° phase angle.

where $a = g/\tan|\alpha|$, α is the phase angle (i.e., the angle included between the incident and scattering directions), and g is the compaction parameter.*

The width of the opposition effect is a function of Hapke's compaction parameter g,

$$g = 2(f)^{2/3} \tag{VII-3}$$

where f is the fractional volume occupied by solid matter.

Table VII-4 presents a calculation of the opposition effect from the Hapke model (referenced to a 30° phase angle) as a function of the fractional volume occupied by solid matter (f). For $MgCO_3$ and incandescent illumination, referring to Table VII-1, an f of 50% yields an opposition effect of 1.25, reasonably close to the observed. From a scanning electron microscope image of this same $MgCO_3$, an estimate of 50% is representative (Egan and Hallock, 1969); for the Nextel paints, a value of f of about 30% appears reasonable (Egan and Hilgeman, 1976).

To apply Hapke's theory, the particles need not be independently suspended in depth. The opposition effect would be expected to be wavelength independent (as observed, Table VII-1) as long as the particles were large compared to the wavelength so that diffraction effects were minimal.

The laser opposition effect (Table VII-3) is similar to that found for less-coherent radiation, except the sulfur is a bit higher; the black and blue paints are very high. The high absorption of the black and blue paints could produce a lower multiply scattered radiation component and a sharper opposition effect. However, this does not explain the large magnitude, which possibly could be caused by interference effects.

Despite the fact that Hapke's theory (1963) predicts reasonable values for the opposition effect, several difficulties are apparent. Specifically, experi-

* The quantity α (phase angle) should not be confused with previous usage as the absorption coefficient.

mental values occur greater than those theoretically predicted. Also, one cannot have shadowing in a nonabsorbing material such as $MgCO_3$; in effect Hapke's theoretical representation violates conservation of energy.

2. Mie Theory Backscatter

As indicated earlier in this chapter, the interesting and puzzling retroreflectance effect of surfaces has an analogy in the "glory" appearing around the shadow of an airplane on a cloud when viewed in the retroreflecting direction. van de Hulst (1957) discusses glory as an application of the Mie theory. A factor of 2 enhancement was all that was necessary to explain previous surface scattering observations such as from sunlight (with a collimation angle of 0.5°). The Mie theory already has been applied to surface scattering with some modifications, and might be applied in the theory of the opposition effect from surfaces.

In order to apply Mie scattering, the Dave (1972) Program I may be used to calculate the intensity of the backscattered radiation at 0.05° intervals from 172 to 180°. For a representative real index of refraction of 1.700 and absorption of 10^{-2}, 10^{-3}, 10^{-4}, and 10^{-5}, the plot of intensity versus scattering angle is shown in Fig. VII-7 for a size parameter of 160. A distinct backscattering peak occurs at 180° for the lower absorbing materials. The retroreflectance may reach 1000 at 180° relative to 178°. While this is extreme, it illustrates a possibility. It is interesting to note that the highly absorbing case ($\kappa_0 = 10^{-2}$) produces no retroreflectance.

The main difficulty with the Mie scattering approach is that it is not yet fully developed. It appears promising and capable of explaining the large values of retroreflection found for some nonabsorbing materials.

3. Lens Hypothesis

More recently, Venable *et al.* (1977) suggested that crystalline materials such as $BaSO_4$ could form a few cube corners and strongly retroreflect; another possibility indicated is that spherical particles may act as lenses. Although these spherical particles then follow the characteristics of Mie scattering, the fact that they are located above a reflecting surface serves to augment the normal Mie backscatter and gives rise to a strong opposition effect.

These hypotheses have not been documented experimentally, but yet they are additionally tenable as explanations for retroreflectance.

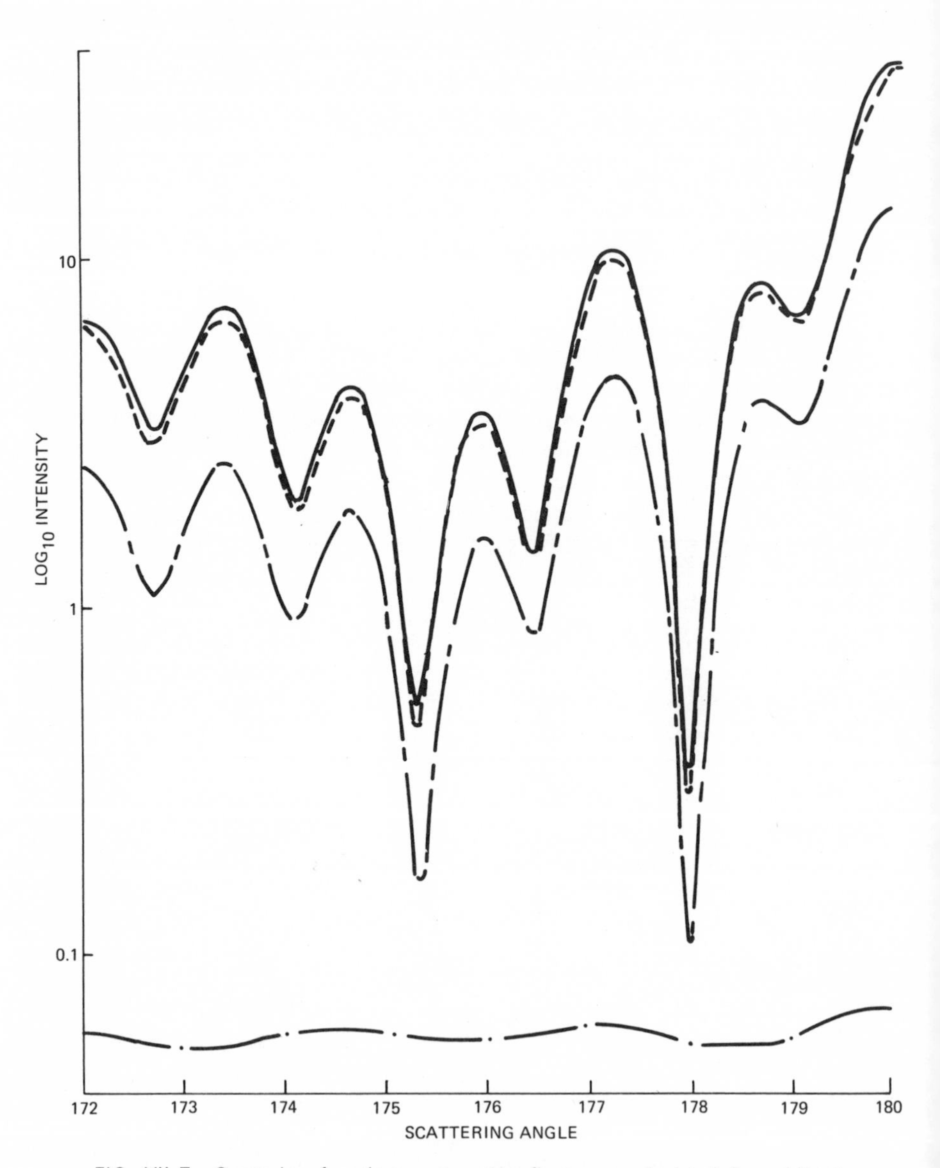

FIG. VII-7 Scattering functions near retroreflectance calculated from Mie theory using a size parameter of 160, a real index of 1.700, and imaginary indices of 10^{-2} —·—, 10^{-3} ——–, 10^{-4} ---, and 10^{-5} ——.

By analogy to glory, rougher particles would exhibit a lower opposition effect. Thus $MgCO_3$ (which has a platelet structure) would have a lower opposition effect than white Nextel paint (which has spherical pigment particles), as observed. Viewing angle would have little or no effect nor would there be any specularity effect, as observed. The large retroreflectance of the sulfur with the laser illumination and smaller collimation–sensor acceptance angle could be the result of its "fairy castle" structure.

Darker surfaces would be more difficult to analyze because of the competing effects of scattering and absorption in the binder and pigment.

NOTES ON SUPPLEMENTARY READING MATERIAL

Mie (1908): Pioneering paper with mathematical description of the scattering of electromagnetic radiation by a sphere.

McLaughlin (1961): A good introduction to astronomy presenting facts, philosophy, history, and related science; a strong emphasis on astrophysics.

Van de Hulst (1957): Mathematical exposition of Mie scattering with a discussion of the backscatter of a raindrop producing a rainbow. (See p. 228 ff.)

CHAPTER VIII □ REMOTE SENSING OF PLANETARY SURFACES

A. INTRODUCTION

In this chapter, the surface models developed in previous chapters will be applied to large-scale remote sensing of the optical properties of planetary rock and mineral surfaces. The modeling described is not as sophisticated as has been presented earlier in this book, but does represent current thinking in remote sensing. Terrestrial remote sensing is excluded because natural ground cover generally confuses the detection of the pristine rock or soil surface; further, atmospheric interferences such as clouds and haze complicate observations. Therefore, the optically most accessible surfaces are those of planets and their satellites, and asteroids; these astronomical objects are viewable by telescope and man-made satellites. The wavelength range that will be considered extends from the ultraviolet (0.185 μm) to the infrared (~ 2.6 μm).

The lunar surface will first be discussed, both from the viewpoint of modeling using the optical complex index of refraction of terrestrial analogs, and from the viewpoint of observations of the large-scale optical effects including the opposition effect and the scattering of radiation as a function of the viewing direction. Then, the surface of Mars will be considered as inferred from the composition of the aerosol dust in the Martian atmosphere. Following this, the rings of Saturn will be described optically and a comparison will be made between two possible predictions of their composition. A brief discussion of the Bruderheim meteorite photometric properties will conclude the chapter.

The principles described permit the extension of the concepts in this chapter to planetary satellites, and asteroids.

B. THE LUNAR SURFACE

A major motivation for the use of the optical properties of the moon's surface in order to characterize it came with the Lunar Landing Program. As millions of dollars were being spent to design the Ranger landers and ultimately the Lunar Module, the designers required definitive information about the lunar surface. That is: Would the lunar surface support a lander?; Was the lunar surface composed of dust that would cover a lander?; How big would the landing pads (feet) have to be so that a lander would not sink into the surface too far?; Could people walk on the surface?; How cohesive was the surface?; and so on.

Thus began the intensive lunar-surface definition program over a decade ago. The lunar surface had been studied extensively before that time, particularly by Lyot (1929) who made polarimetric observations, followed by Dollfus (1957, 1961). Photometric observations were made by van Diggelen (1959, 1965). Most researchers made measurements at one wavelength and used either polarimetry or photometry. Gehrels *et al.* (1964) broke this tradition and used both polarimetry and photometry in a range of wavelengths. At the time, the astronomical techniques utilized broadband colored glass filters to isolate spectral wavelength regions. Letters were used to code the effective wavelength regions of the filters as shown in Table VIII-1.

TABLE VIII-1 Code for Effective Wavelength Regions of Filters

Code letter	N	U	B	G	V	R	I
Effective wavelength (μm)	0.33	0.36	0.44	0.54	0.56	0.69	1.06

Even with the advent of precision high-resolution interference filters, this system still remains in use, partly because of the extensive amount of data acquired under it and partly because the original selection of wavelengths corresponds to important stellar features. This section will discuss only the photometric properties of the lunar surface, polarimetric surface modeling still being in an early stage.

A description of the lunar surface involves the use of optical complex indices of refraction of lunar samples or analogs and geometrical factors such as the interparticulate shadowing of the particles or structures constituting the surface to predict retroreflectance and other phenomena. The optics of the lunar surface will be introduced using large-scale astronomical observational data. The interpretation of photometric behavior involves the use of various mathematical functions that represent the effects of interparticulate shadowing as the incident radiation and viewing directions change. A number of approaches to lunar surface modeling have been presented in the literature, and these will be discussed in this section, along with large-scale laboratory photometric measurements of terrestrial analogs. Then the lunar surface will be characterized using the optical complex indices of refraction of terrestrial analogs, and a comparison will be made to the analyses of the lunar surface.

Representative photometric reflectance curves are presented in Fig. VIII-1; the brightness in the V spectral region (Table VIII-1) is plotted versus phase angle (Gehrels *et al.*, 1968). This terminology requires some elucidation; the unit of brightness used here is 3.90 ($\pm$0.05) $\times 10^{-9}$ erg cm^{-2} sec^{-1} Å^{-1} for a region 5″.02 apparent diameter on the moon at 1 astronomical unit (1 a.u.) from the sun, and the phase angle is the sum of the incident angle and viewing angle when they are coplanar. The reason for the very specific brightness unit is that certain standard stars are used as references in astronomical brightness measurements. This procedure is followed because the transmission of the optics of the particular telescope is uncertain, and the earth's atmosphere varies. Thus, universal stellar references are usually employed for calibration. The phase angle, used for the abscissa, is the angle between the incident and scattered radiation from a surface. The plane of the phase angle need not contain the normal to the surface. It has been observationally found that the photometric function for the lunar surface depends mainly upon the phase angle, and is nearly independent of the location on the lunar sphere.

Figure VIII-1 shows a lunar surface brightness that increases with decrease in phase angle, and a nonlinear upsurge near 0° phase angle (the opposition effect). A perfect diffuser (cosine surface) reflects as shown by the dashed curve in the lower portion of the figure. Different lunar regions have differing photometric functions depending upon the surface structure and mineralogy. The gradual increase in brightness with decrease in phase angle is caused by adjacent surface particles shadowing each other, and will be described further in the following. The opposition effect was covered in Chapter VII.

The fundamental references for the shadowing effect are not readily accessible (Seeliger, 1887; Bobrov, 1940, 1963; Franklin, 1962). Hapke (1963) independently derived the theory, but a number of flaws existed in the analysis, which was later improved by Irvine (1966).

Hapke (1963) postulated a photometric function for a particulate medium. For specific lunar regions, it was

$$\Phi(\alpha, \lambda) = \frac{I(\alpha, \lambda)}{I(0)} = \frac{\sin|\alpha| + (\pi - |\alpha|)\cos|\alpha|}{\pi + \pi\cos\lambda/\cos(\lambda + \alpha)} B(\alpha, g) \qquad \text{(VIII-1)}$$

where Φ is the photometric function (normalized to 1 at $\alpha = 0°$), α the phase angle, λ the selenographic longitude relative to the subearth point, $B(\alpha, g)$ the opposition effect function [see Eq. (VII-1)], $I(0)$ the observed intensity

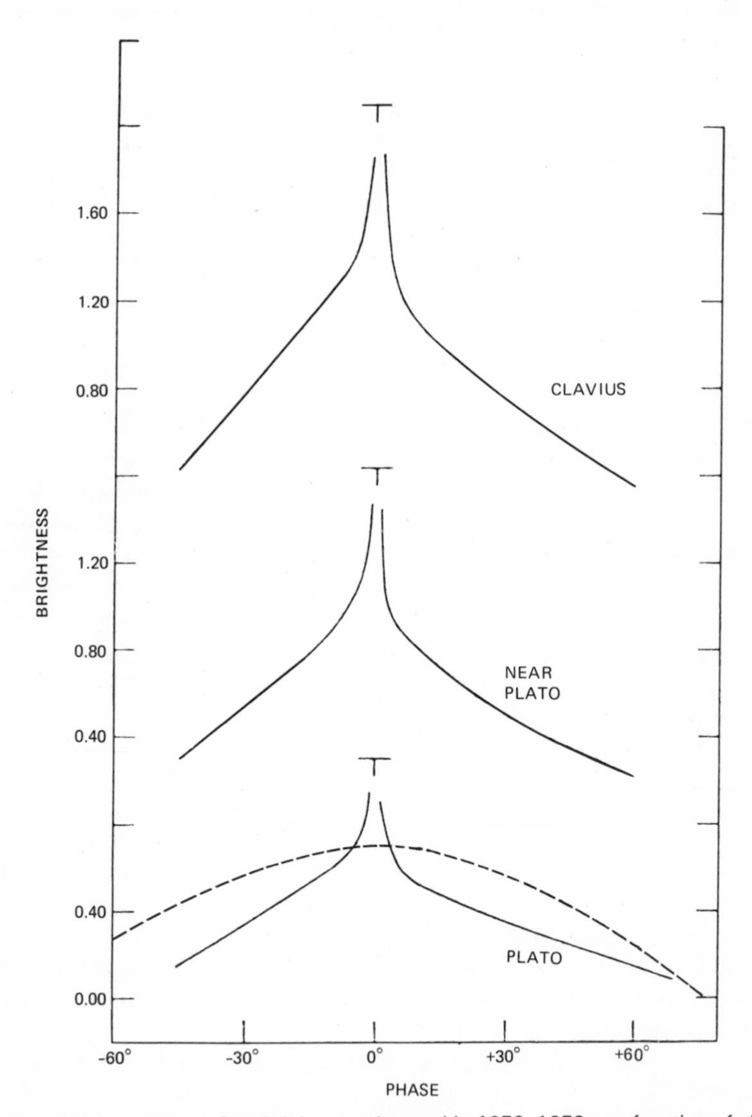

FIG. VIII-1 Lunar surface brightness observed in 1956–1959 as a function of phase; the ordinate represents the brightness in the V spectral region (effective wavelength = 0.56 μm); the T symbols represent the extrapolated brightness at zero phase angle. The dashed curve represents a Lambertian surface. [From Gehrels *et al.* (1964).]

at $\alpha = 0°$ [Gehrels *et al.* (1964) indicated that $I(1.5°, \lambda) \cong I(0)$], $I(\alpha, \lambda)$ the observed intensity as function of α and λ, and $g \cong 2$(fractional volume occupied by solid matter)$^{2/3}$.

Fairly good agreement was obtained with the experimental observations with this expression. Later Hapke (1966) improved the theoretical photometric function to have three independent parameters f_0, γ_0, and g. He derived an improved Lommel–Seeliger law as

$$L(\alpha, \lambda) = \left\{K_1 \frac{1 - f_0}{1 + \cos\lambda/\cos(\lambda + \alpha)} + K_2 \frac{f_0}{2\cos\frac{1}{2}\alpha\cos\lambda\sin\gamma_0} \times \left[\cos(\lambda + j\alpha)\sin(\gamma_0 + k\alpha) - \frac{1}{2}\sin^2\frac{1}{2}\alpha\ln\left|\frac{\cos(\lambda + j\alpha) + \sin(\gamma_0 + k\alpha)}{\cos(\lambda + j\alpha) - \sin(\gamma_0 + k\alpha)}\right|\right]\right\} \tag{VIII-2}$$

where K_1, K_2, j, and k have different values, depending upon α and λ, and are given in Table VIII-2, and α is the phase angle, λ the selenographic longitude relative to subearth point, f_0 the fraction of surface occupied by cylindrical depressions, and γ_0 the half angle of depression. In order to relate $L(\alpha, \lambda)$ to $\Phi(\alpha, \lambda)$, one must account for the additional angular dependences of the opposition effect $B(\alpha, g)$ and $\Sigma(\alpha)$, which accounts for the spherical shape of the moon, i.e.,

$$\Phi(\alpha, g, \lambda) = L(\alpha, \lambda)\Sigma(\alpha)B(\alpha, g) \tag{VIII-3}$$

and

$$\Sigma(\alpha) = [\sin|\alpha| + (\pi - |\alpha|)\cos|\alpha|]/\pi + 0.1(1 - \cos|\alpha|)^2 \tag{VIII-4}$$

TABLE VIII-2 Constants for Use with Eq. (VIII-2)[a,b]

Region	Definition	K_1	K_2	j	k
0	$-\frac{1}{2}\pi \le \lambda \le \frac{1}{2}\pi, \frac{1}{2}\pi - \lambda \le \alpha \le \pi$	0	0	—	—
1	$-\frac{1}{2}\pi + \gamma \le \lambda \le \frac{1}{2}\pi, \frac{1}{2}\pi - \gamma - \lambda \le \alpha \le \frac{1}{2}\pi - \lambda$	1	1	1	$\frac{1}{2}$
2	$-\frac{1}{2}\pi + \gamma \le \lambda \le \frac{1}{2}\pi - \gamma, 0 \le \alpha \le \frac{1}{2}\pi - \gamma - \lambda$	1	1	$\frac{1}{2}$	0
3	$-\frac{1}{2}\pi \le \lambda \le -\frac{1}{2}\pi + \gamma, 0 \le \alpha \le \frac{1}{2}\pi - \gamma - \lambda$	1	1	0	$\frac{1}{2}$
4	$-\frac{1}{2}\pi \le \lambda \le -\frac{1}{2}\pi + \gamma, \frac{1}{2}\pi - \gamma - \lambda \le \alpha \le \pi - \gamma$	1	1	$\frac{1}{2}$	1
5	$-\frac{1}{2}\pi \le \lambda \le -\frac{1}{2}\pi + \gamma, \pi - \gamma \le \alpha \le \frac{1}{2}\pi - \lambda$	1	0	—	—

[a] From Hapke (1966).

[b] Only values appropriate to $\alpha \ge 0$ are given explicitly. The photometric function is symmetric with respect to a simultaneous change of sign of α and λ.

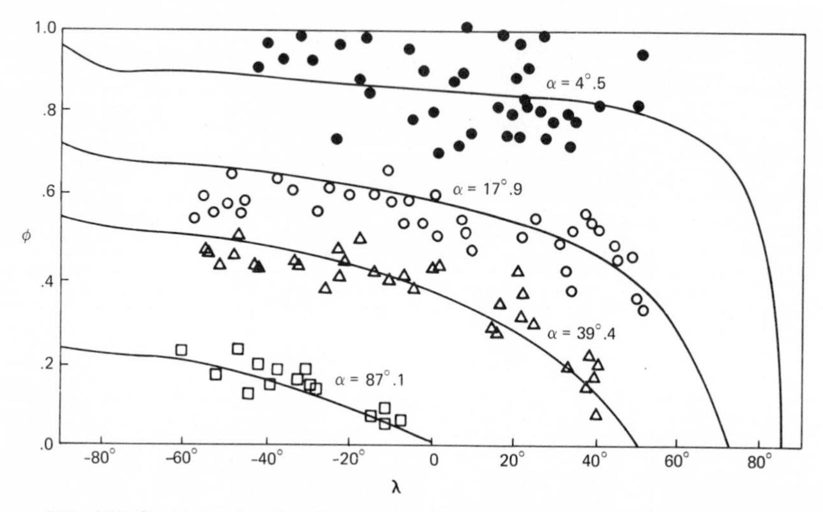

FIG. VIII-2 Normalized brightness Φ of lunar surface areas as a function of longitude for several phase angles. Solid lines, theoretical photometric function of Hapke (1966); points, data of Fedoretz [after Parker *et al.* (1964)]. [From Hapke (1966).]

The effectiveness of this expression is illustrated by the solid lines in Fig. VIII-2, which depict the normalized brightness of lunar surface areas as a function of lunar longitude for several phase angles in comparison to the data points of Fedoretz, as corrected for normal albedo by Parker *et al.* (1964).

The discussion of the shadowing effect in diffuse reflection by Irvine (1966) incorporates a review of results in the astronomical literature; the correction for the shadowing effect is inserted into the radiative transfer theory in terms of integrals, and not in the explicit form of Hapke (1963, 1966).

Other attempts at representing shadowing have been made in the past; Barabashev (1922) used grooves and clefts, Shönberg (1925) considered spherical domes, and Bennett (1938) and van Diggelen (1959) employed hemiellipsoidal cups. A novel approach to theoretically representing shadowing has been suggested by Halajian and Spagnolo (1966), using macroscopic structures (Fig. VIII-3) to account for the lunar phase curve. They started with suspended horizontal nontransparent strips to produce a contrived model, and then added similar vertical strips. These were combined into the "T" model (Fig. VIII-3a) as the model became more elaborate. The matching was good at normal viewing, but deteriorated at larger phase angles; the addition of lips at the tops of the "T" structures (modified "T" model; Fig. VIII-3b) produced additional shadowing at larger phase angles, and a better match. The addition of lips produces a rudimentary version of a vesicular structure (Fig. VIII-3c), whereby a surface porosity

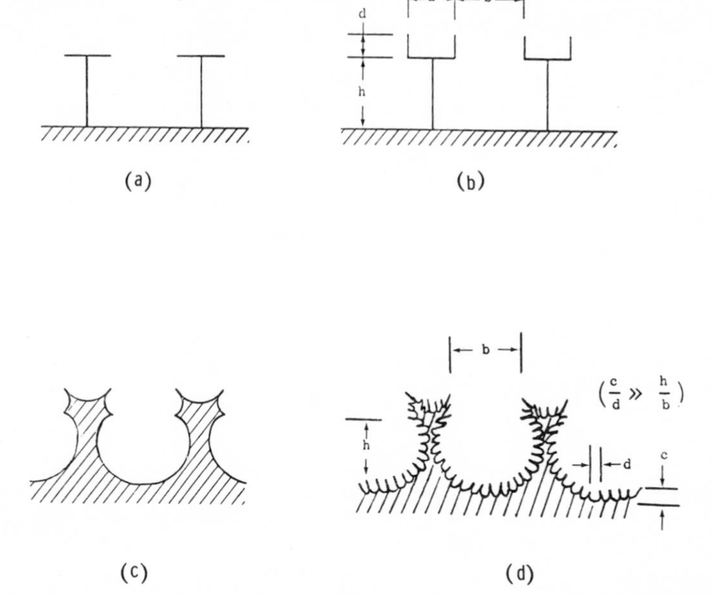

FIG. VIII-3 Evolution of building block for a lunar photometric model (a) Basic T model, (b) modified T model (to improve backscatter), (c) cellular model (evolved from modified T model), and (d) modified cellular model (to account for opposition effect). [From Halajian and Spagnolo (1966).]

can be estimated for a Hapke-type surface model. The opposition effect is not produced by these models, and the modified cellular model (Fig. VIII-3d) was suggested as adding enough microstructure to produce the opposition effect. The matching is illustrated in Fig. VIII-4. The solid line represents an average lunar photometric curve that has been truncated at a 5° phase angle. The dashed line represents the matching for the modified "T" model.

Many attempts have been made to match the observed photometric properties of the lunar surface with laboratory models. Hapke and van Horn (1963) used solid rocks, volcanic slags, and coarsely and finely ground rock powders in an attempt to match the lunar photometric curve. Finely divided dark powders such as silver chloride (Fig. VIII-5), nickel sulfate, and cupric oxide were sifted to form complex surfaces of low density which matched the lunar curve (Figs. VIII-4 and VIII-6) very well. Hapke and van Horn (1963) inferred that the surface of the moon is covered with 10-μm-average-diameter low-albedo particles with 90% of the volume being voids.

In another approach, Halajian and Spagnolo (1966) were able to match a lunar surface with Hawaiian volcanic cinder (Fig. VIII-6). An average lunar surface is shown shaded with a range of lunar curves normalized to

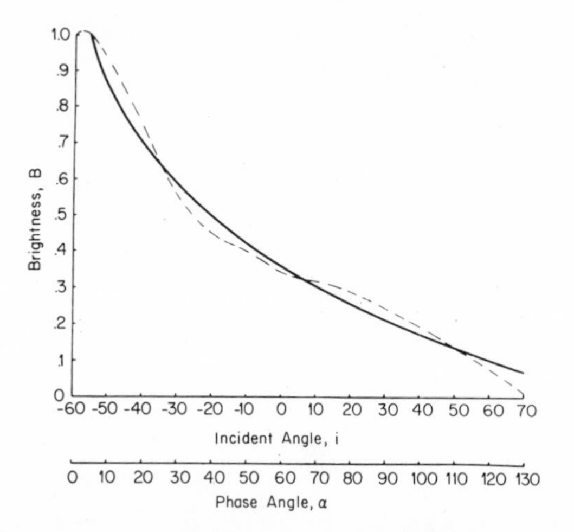

FIG. VIII-4 Photometry of modified T model for uniform albedo Lambert scattering (observing angle = 60°), $a = 1.0$, $b = 1.6$, $h = 1.0$, and $d = 0.2$; solid curve is average lunar curve observed, dashed is calculated. [From Halajian and Spagnolo (1966).]

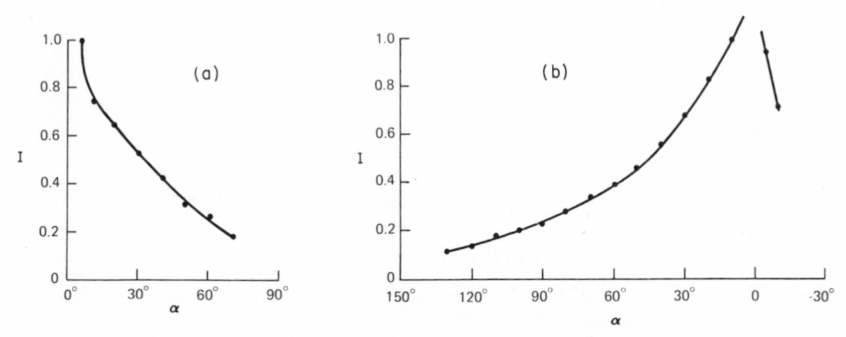

FIG. VIII-5 Phase function of silver chloride powder, albedo = 0.10; (a) viewing angle = 0°, (b) viewing angle = 60°. (Hapke and van Horn, *Journal of Geophysical Research*, **68**, 4545, 1963. Copyrighted by American Geophysical Union.)

1.0 at 5° phase angle for a viewing angle of 60°. The solid curve is the photometric function measured for a Hawaiian volcanic cinder with an albedo of 0.11.

From these observations, it can be seen that neither the mathematical representation nor the physical configuration is unique.

An analysis of the lunar surface in terms of the optical complex index of refraction will now be discussed.

Some measurements of the optical complex index of refraction have been reported on selected lunar materials such as ilmenites (Simpson and Bowie,

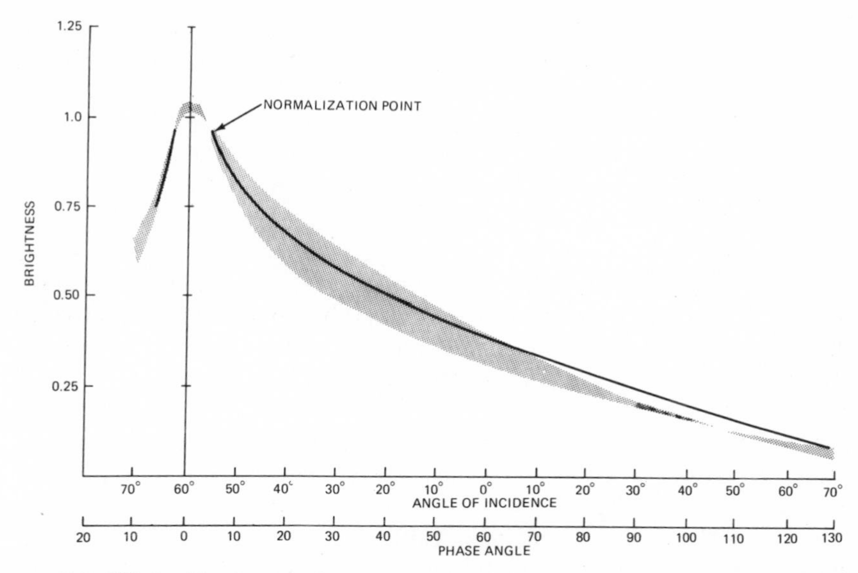

FIG. VIII-6 Photometry of Hawaiian volcanic cinder measured with beam splitter, incident angle of 60°; shaded area is average lunar surface, solid line is for sample with albedo = 0.11. [Adapted from Halajian and Spagnolo (1966).]

1970) in the visual range, pyroxenes in yellow light (Hargraves *et al.*, 1970) of the real portion only, and "absorbance" or reflectance observations (Adams and Jones, 1970; Birkebak *et al.*, 1970; Bell and Mao, 1972) that have only an imprecise correlation with the absorption coefficient of the optical complex index of refraction. Representative terrestrial analogs of dominant lunar minerals are a plagioclase feldspar–bytownite (Minnesota) with approximately 20% $NaAlSi_3O_8$ and about 80% $CaAl_2Si_2O_8$, and a clinopyroxene–augite (Canada), $(Ca, Na)(MgFe^{2+}Fe^{3+}, Al)(Si, Al)_2O_6$. These minerals, both silicates, represent on the average about 25 and 50%, respectively, of the lunar surface (Weill *et al.*, 1970). These terrestrial minerals are only an approximation for lunar rock, because other elements, such as titanium in ilmenite, not present in the terrestrial analogs, would be expected to cause some differences in the complex index of refraction.

The real portion of the optical index of refraction of bytownite (Chapter V) decreases from the ultraviolet to the infrared, with minima at 1.4 and 2.0 μm. The real index of augite (Chapter V) also similarly decreases with a leveling off at 0.8 μm. The values for the real indices are closely comparable to the accepted values for single crystals of the minerals in visual light [bytownite: $n_\alpha = 1.564$ to 1.573, $n_\beta = 1.569$ to 1.579, and $n_\gamma = 1.573$ to 1.585; augite: $n_\alpha = 1.688$ to 1.712, $n_\beta = 1.701$ to 1.717, $n_\gamma = 1.713$ to 1.737 (Kerr, 1959)]. However, the real index, in yellow light, as measured by other investigators, for lunar pyroxenes is slightly higher, varying between 1.702 to 1.755 (Hargraves *et al.*, 1970).

The results of total transmission and reflectance measurements for the determination of the imaginary (absorption) portion of the complex index of refraction for bytownite were presented in Chapter V for the 0.0254-cm thin section. The total transmission is relatively high with minima at 0.3, 0.6, 1.2, and 2.0 μm. The total reflection spectrum has similar features and, thus, can be used conveniently in a comparison to observed lunar-sample reflectance spectra. Ultimately, the most desirable approach would be to use the optical complex indices of refraction in a Monte Carlo surface model, for instance, and compare the calculated with the actual lunar-sample spectra.

The absorption band at 1.2 μm has been attributed to electronic transitions, probably due to ferrous ions in a somewhat disordered site (Hunt and Salisbury, 1970), although analyses of the sample reveal very little iron (Egan and Hilgeman, 1975b; also Table V-2). Adams and Filice (1967) indicate that bands at 1.4 and 1.9 μm in crystalline acidic rocks, such as rhyolite, are characteristic; rhyolite is composed essentially of alkali feldspar, of which bytownite is representative, and quartz. Adams and Filice (1967) suggest that bands at 1.1 μm would be due to Fe^{2+}, with shifts possibly caused by differences in metal–oxygen distances. They suggest that the band at 1.9 μm, which appears in most crystalline acidic rocks, may also be due to Fe^{2+} on silicon sites in tetrahedral coordination. The presence of both 1.4- and 1.9-μm bands usually indicates water molecules; broad bands are indicative of relatively unordered sites, and/or that more than one site is occupied by water molecules (Hunt and Salisbury, 1970). The absorption band at 0.6 μm in the bulk sample is not observed by Hunt and Salisbury (1970) on their bytownite (sample 106B), although a hint of it is indicated in the flattening out of the 250–1200-μm powder curve at 0.6 μm relative to the finer powders; pulverization tends to wash out absorption bands. It has been suggested that a 0.6-μm band would be due to Fe^{2+} or Fe^{3+} (Hunt and Salisbury, 1970) or to Ti^{3+} (McCord *et al.*, 1970), or to FeS (Egan and Hilgeman, 1972). Iron is observed only in trace quantities in the sample, and titanium not at all.

The bytownite absorption band at 0.3 μm is attributed to electronic transitions in the metal atoms (Hunt and Salisbury, 1970).

The transmission of the 0.010-cm section of the clinopyroxene, augite, is presented in Chapter V. After allowing for the difference in thickness of the samples, the transmission of the augite is about three orders of magnitude less than bytownite in the ultraviolet, and an order of magnitude less in the infrared.

Weak reflection minima occur in the ultraviolet and at the 0.6- and 1.2-μm bands, attributed to Fe^{2+} by Adams and Jones (1970) in the lunar clinopyroxene. These do not appear in the terrestrial clinopyroxene sample that contains 6% iron, but they do appear in the bytownite that has ~0.01% iron. The bands at 0.7 and 1.2 μm in the sample are not the result of TiO_2, suggested as being the cause of the corresponding bands in Apollo 11 lunar samples (Adams and Jones, 1970), since there is no detectable titanium in the sample. Note that the bytownite appears to have these bands (2.0, 1.2, and 0.6 μm), and it is possible that they could thus be representative of the plagioclase feldspar in the lunar rock. The high slope of the region enveloping the lunar 0.5-μm band (Adams and Jones, 1970) would shift an absorption minimum occurring at a larger wavelength to a shorter wavelength.

The bytownite absorption bands correspond closely in wavelength to those observed in reflection, but not completely so for the more highly absorbing augite. For an augite thin section (that visually appears greenish in transmitted light), the peak absorption occurs at 0.7 μm, whereas the minimum total reflectance occurs at 0.6 μm. The discrepancy appears to be the result of the relatively high transmission at 0.6 μm.

As a result of the measurements of the optical complex indices of refraction of representative terrestrial analogs of the dominant lunar constituents, plagioclase feldspar (bytownite), and pyroxene (augite), comparable absorption peaks to those obtained from reflectance measurements were found at 2.0, 1.3, 1.0–0.9, 0.6 μm, and in the UV. The similarities indicate that the optical constants presented reasonably characterize the lunar surface material.

Thus, one could reasonably use the optical complex index of refraction of these constituents, in an appropriate surface model, to represent the diffuse scattering properties of the lunar surface. The opposition effect is not yet amenable to an exact theoretical treatment.

C. THE MARTIAN SURFACE

There is considerable interest in the composition of the aerosols in the atmosphere of Mars as indicators of the surface material. A knowledge of the aerosol optical properties can be inferred, for example, from the interpretation of ultraviolet photometric data from space probes.

The present gaseous atmosphere of Mars is composed primarily of carbon dioxide with a mean surface pressure of 5.5 mbar (Woiceshyn, 1974). There appear to be aerosol constituents, with augmentation by planetary surface material during dust storms. Since very fine dust particles produce stronger optical scattering as the wavelength decreases into the ultraviolet, Mars atmospheric-aerosol compositional information may be inferred from ultraviolet observations. One good opportunity occurred early during the Mariner 9 reconnaissance of Mars in 1971, when a dust storm occurred which completely obscured the planet's surface. Observational data exists in the ultraviolet wavelength range between 0.20 and 0.35 μm (Pang and Hord, 1973; Pang *et al.*, 1976).

A number of possible materials have been suggested as constituting the aerosols in the Martian atmosphere. One of the first suggestions was limonite (Dollfus, 1961). Although the limonite classification covers a broad range of minerals, one sample from Venango County, Pennsylvania, was included. The sample has been described previously (Egan, 1969, 1971).

Subsequently, as a result of infrared interferometer spectrometer measurements of the Martian dust storm from the Mariner 9, Hanel *et al.* (1972) proposed that a $60 \pm 10\%$ SiO_2 (by weight) material would be the most probable for the dust, suggesting andesine or oligoclase. Therefore, andesite, composed of andesine and oligoclase, might be an appropriate rock (Lowman, 1974). Of many samples examined, one brownish olivine–pyroxene andesite from Volcano Tunaba, 70 miles southwest of Limon, Costa Rica, was measured (Columbia University Petrographic Collection No. 1.2.1.2.5).

Hunt *et al.* (1973) found that montmorillonite, an alteration product described by the general empirical formula

$$(\tfrac{1}{2}Ca, Na)_{0.7}(Al, Mg, Fe)_4(Si, Al)_8O_{20}(OH)_4 \cdot nH_2O,$$

exhibited spectral transmission properties that closely matched the available Martian data between 6 and 12 μm. A sample of an Amory, Mississippi brownish montmorillonite was obtained from the Lunar and Planetary Branch, Air Force Cambridge Research Laboratories; the sample has been described by Hunt and Salisbury (1970) and is designated as montmorillonite 222B in their paper. A basalt sample from Chimney Rock, New Jersey, Watchung Mountain Range (initially measured by Egan and Becker, 1968, 1969) could be included also. Adams and McCord (1969) have suggested oxidized basalt as a possible Martian surface material.

Visually, the limonite and brown montmorillonite samples approximate the color of Mars. However, the andesite is not as good a match, and the basalt, being gray, is the poorest match.

Using the optical complex indices of refraction for these materials (Chapter V), a comparison may be made with those inferred from the Mariner 9 ultraviolet spectra of the 1971 Mars dust storm. Pang *et al.* (1976) matched the observed single-scattering albedo and phase function with Mie scattering calculations for a range of size distributions for spheres of homogeneous and isotropic material (Fig. VIII-7). The real component of the index

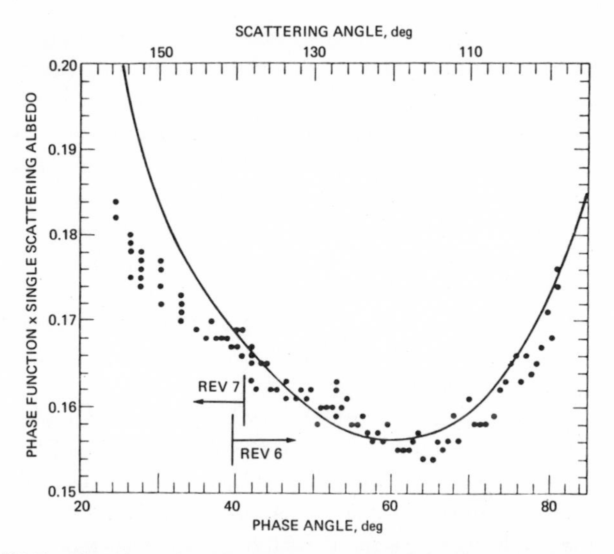

FIG. VIII-7 The phase function observed in a 10-nm band centered at $\lambda = 305$ nm during revolutions 6 and 7 of Mariner 9 is plotted as dots. The matching theoretical curve was calculated with a size distribution of spheres of effective radius $a = 1.5$ μm, effective variance $b = 0.2$, complex refractive index $\hat{m} = 2.2 - 0.01i$, and $F = 0.75$. [From Pang *et al.* (1976).]

of refraction was inferred to be $\gtrsim 1.8$ at both 0.268 and 0.305 μm; the imaginary component was inferred to be 0.02 at 0.268 and 0.01 at 0.305 μm. None of the terrestrial analogs considered had exactly the required ultraviolet properties, the refractive indices being too low and the absorption decreasing too slowly between 0.200 and 0.400 μm. Thus, appropriate samples have not yet been found to represent the dust.

D. THE RINGS OF SATURN

The exact composition and structure of the rings of Saturn has not yet been determined. Although the rings of Saturn probably contain H_2O ice in some form [based on the Kuiper *et al.* (1970) observations and the Pilcher *et al.* (1970) analyses], the associated minerals are open to question. Observations of the spectral reflectivity of the rings by Lebofsky *et al.* (1970) in the spectral range from 0.3 to 1.05 μm show a sharp drop in the reflectivity from 0.6 to 0.3 μm. Water ice is very transparent in this region, so that this feature must be attributed to the presence of some other material in the rings; the observation cannot be explained by having some particles made solely of absorbers (such as "silicates") and others made up solely of H_2O ice (Pollack *et al.*, 1973). Unfortunately, in determining this ultraviolet-absorbing material, compositional remote sensing based on reflectance comparisons alone is fraught with pitfalls. This is because particle geometry can have a strong effect on the reflectance (Emslie and Aronson, 1973; Aronson and Emslie, 1973; Egan and Hilgeman, 1978a). In fact, phase angle and polarization studies often are used to provide information on structure (Price, 1973; Cook *et al.*, 1973; Dollfus, 1961).

Compositions suggested for the rings range from ice with an admixture of dust, or clathrate hydrates (Cook *et al.*, 1973), and silicate–water ice particles (Pollack *et al.*, 1973), to an ultraviolet-irradiated H_2O frost contaminated by H_2S (Lebofsky, 1973; Wamsteker, 1975).

In a quest for possible candidate materials having such a sharp drop in reflectance between 0.6 and 0.3 μm, the concepts of Lewis (1974) concerning the chemistry of the solar system may be applied. One could suppose that the rings of Saturn could be composed, in part, of core material hypothesized for the outer planets. The core material would be an iron–nickel alloy if the planetary system was formed by inhomogeneous accretion, or an iron sulfide such as troilite if the formation occurred through equilibrium condensation (Lewis, 1974). Since iron and nickel reflectances are not appropriate between 0.3 and 0.6 μm (Hodgman, 1950) for the rings of Saturn, a probable candidate material to be considered for the rings would be troilite. Troilite is stoichiometric FeS; a related mineral is pyrrhotite, which has a deficiency of iron of up to 20% (Hurlbut, 1959).

The deducing of the surface composition of planets by reflectance measurements is modified by the effect of optical scattering of the incident radiation in the surface material of the planet. The optical scattering is a function of the optical complex index of refraction of the surface material as well as geometrical factors such as particle size, shape, and packing. The fundamental optical property—the optical complex index of refraction as a function of wavelength—when used in an optical model of the planetary surface, can reduce the indeterminacy of the knowledge of the composition of the surface (Emslie and Aronson, 1973; Aronson and Emslie, 1973; Egan and Hilgeman, 1978a).

Using the optical complex indices of refraction given in Chapter V, the normal reflectivity R_N of troilite and pyrrhotite as a function of wavelength is computed and is shown graphically in Fig. VIII-8, normalized to the ring-*B* data at 0.700 μm of Lebofsky *et al.* (1970). The ultraviolet–blue–visible (UBV) data of Franklin and Cook (1965) are also shown, but these are for other ring tilts. There also has been noted an east–west spectrophotometric asymmetry (Kozyrev, 1974). Neither the troilite nor the pyrrhotite match the ring-*B* data of Lebofsky *et al.* (1970) exactly, but

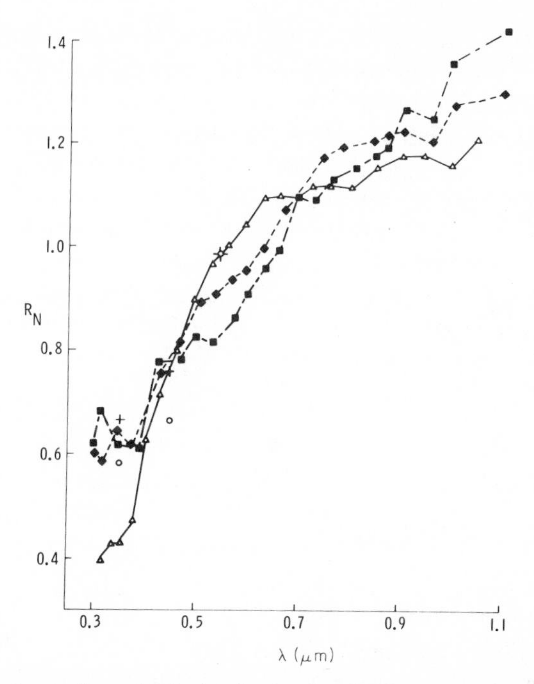

FIG. VIII-8 Comparison of calculated normalized spectral reflectivity R_N of pyrrhotite (--- and ◆) and troilite (——— and ■) with the observations of the rings of Saturn made by Lebofsky *et al.* (1970) (△) and the UBV observations of Franklin and Cook (1965); +, post opposition; ○, preopposition. [From Egan and Hilgeman (1977b).]

the trends are what are significant; essentially, the pyrrhotite matches best between 0.4 and 0.7 μm, coming closer to the Franklin and Cook (1965) data near 0.35 μm. The troilite is a poorer match in this region. But for both semiconductors (troilite and pyrrhotite), there is a matching problem at wavelengths larger than 0.7 μm; the curves must be depressed.

Because ice has been observed in the infrared as a constituent of the rings (Kuiper *et al.*, 1970), and some silicate content is also likely based on an equilibrium condensation model of planetary formation (Lewis, 1974), we must consider the effect of these materials. For instance, a better fit to the data of Lebofsky *et al.* (1970) could be accomplished with the addition of a small quantity of H_2O ice [having an absorption band at 1.0 μm (Irvine and Pollack, 1968)] and of enstatite, an orthopyroxene, which has a strong absorption in this region (see Fig. V-10 and Table V-12). These materials have the required absorption that would produce the desired depression in the infrared. It is to be noted that the region near 0.3 μm would be lowered by the enstatite. Enstatite is used as an example of a silicate because of the requirements of the equilibrium condensation model; however, other orthopyroxenes could be suitable. A ferromagnesium silicate is one of the materials occurring in Lewis's (1974) equilibrium condensation model of the solar system. The "enstatite" (Egan and Hilgeman, 1975b) is partially modified, but not as strongly as Lewis's model requires. The reflection model used is simplistic, and there admittedly are other possibilities. The materials chosen are chemically suggested by an equilibrium condensation model for the formation of the solar system. The troilite alone is not consistent with the radar reflectivity and the radio emissivity data. However, a two-component model can be derived which is consistent. The ice, suggested by the near-infrared observational data, is a major factor in explaining the radar reflectivity and radio emissivity data.

E. ASTEROIDS AND METEORS

It has been noted that the 0.6-μm absorption band in asteroids may be related to that found in FeS (Egan *et al.*, 1973b). Note that a 0.6-μm band is observed in the reflection curve of Vesta (McCord *et al.*, 1970). It has also been observed in terrestrial samples of the Bruderheim meteorite (Egan *et al.*, 1973b), as well as in the spectra of other asteroids (Chapman, 1972). An absorption band in this region is also evident in the interstellar extinction curve (Wickramasinghe, 1970).

Photometric and polarimetric laboratory measurements were made as a function of phase angle in the U (0.36 μm), G (0.54 μm), and R (0.67 μm) bands for 0°, 30°, and 60° incident illumination on four particle-size ranges of the Bruderheim meteorite, an *L*6 olivine–hypersthene chondritic meteorite. The four particle-size ranges were 0.25–4.76 mm, 0.25–4.76 mm coated with <74-μm powder, 74–250 μm, and <37 μm. In addition, normal reflectance measurements were made in the spectral range from 0.31 to 1.1 μm. Comparison with astronomical data reveals that none of the asteroids in the main belt for which adequate observations exist can be matched with the Bruderheim meteorite samples, which is representative of the most common meteoritic material encountered by the earth. However, it appears from the photometry data that the surface of the Apollo asteroid Icarus is consistent with an ordinary chondritic composition (Fig. VIII-9). This suggests the possibility that this material, although common in earth-crossing orbits, is rare as a surface constituent in the main asteroid belt.

The optical surface modeling for the Bruderheim meteorite has been described in Chapter VI.

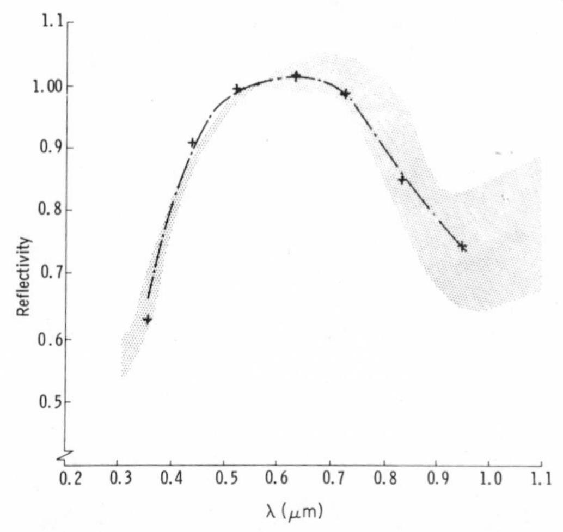

FIG. VIII-9 Comparison of the Bruderheim meteorite spectral reflectance curves with that of the asteroid Icarus; the curves have been normalized to unity at 0.566 μm; Bruderheim range (shaded area), Icarus at 90° phase (—·—) (Data from Gehrels *et al.*, 1970). [From Egan *et al.* (1973b).]

NOTES ON SUPPLEMENTARY READING MATERIAL

Allen (1963): Presents the essential quantitative information of astrophysics; a gathering of information from many references.

Mason (1962): Comprehensive account of meteorites; describes chemical and mineralogical composition as well as structure.

McLaughlin (1961): A good introduction to astronomy; chapter on physical features of the moon and planets.

Wendlandt and Hecht (1966): The two-flux model presented is of historical interest in surface modeling.

CHAPTER IX □ REMOTE SENSING OF THE INTERSTELLAR MEDIUM

The applications of absorption and scattering described in this chapter deal with regions of space beyond our solar system. In particular, the properties of the galactic interstellar medium will be discussed. The interstellar medium is composed of the gas and dust that exists between stars in our galaxy, and it contains a significant fraction (about 10%) of the entire mass of the galaxy. The dust contained in the interstellar medium will be considered in terms of scattering and absorption in the ultraviolet, and inferences made as to possible dominant mineral constituents.

While we now accept the fact that very small particles exist between the stars in our galaxy, it is only recently that a convincing demonstration has been presented of absorption by interstellar dust. Trumpler (1930a, b) first showed that absorption was present, and he estimated the extent. By observing the actual angular diameters of open galactic clusters, together with those calculated from the observed photographic brightness (magnitude) and the known absolute brightness, a discrepancy was found. This difference was the result of interstellar absorption.

Even before this demonstration, absorption was suspected from comparisons of the number of stars brighter than an arbitrarily chosen level in clear regions of the sky (Struve, 1847). However, the accuracy of estimating absorption was severely limited by the uncertainties in the stellar density distribution (Kienle, 1923).

Three-color photoelectric photometry by Stebbins *et al.* (1934, 1939) proved conclusive in the establishment of the existence of the interstellar medium. They measured the ultraviolet–blue–visible (UBV) magnitudes of stars which were in the same color class. They found that the color difference between the blue and visible filters (i.e., B − V) varied with apparent stellar magnitude. The effect was an apparent reddening of the stars with λ^{-1} wavelength dependence.

It is possible that temperature differences of the stars could account for the λ^{-1} dependence, but, with more detailed six-color photometry, the reddening was found to be stronger in the ultraviolet, ruling out this possibility. Scatterers that have been suggested range from metallic particles (Schalen, 1939), to condensed interstellar gas (Lindblad, 1935), H_2O ice (Van de Hulst, 1946, 1949), large molecules with unfilled electronic shells (Platt, 1956), graphite grains (Hoyle and Wickramasinghe, 1962), and silicates (Huffman and Stapp, 1971). As yet, there is no conclusive evidence on the composition of the interstellar medium, and all possibilities remain open. The ultraviolet region has the strongest scattering, and hence is of most interest in investigating interstellar extinction.

Interstellar extinction measurements were extended into the ultraviolet beyond 2200 Å by the Orbiting Astronomical Observatory (Stecher, 1965, 1969; Bless *et al.*, 1968; Bless and Savage, 1970, 1972). A conspicuous feature in many of the extinction curves occurs at about 0.22 μm, apparently varying in shape. Figure IX-1 is a plot of the relative astronomical magnitude of absorption versus the reciprocal of the wavelength measured in micrometers; this convention is used to expand the ultraviolet scale and relate to the conventional astronomical measurement of star brightness in magnitude. This same absorption feature occurs for stars in widely separated galactic regions. Also, interstellar silicate absorption has been observed near 10 μm in the infrared, and this has indicated the existence of silicates in the interstellar medium (Knacke *et al.*, 1969). In addition, absorption features appear at 3.08 μm, and these are attributed to interstellar ices (Merrill *et al.*, 1976).

As an application of the use of the optical complex index of refraction in scattering calculations, and simultaneously as an inquiry into the uniqueness of the present concepts of the composition of the interstellar medium, additional candidate minerals were investigated. They were chosen from those that are highly probable in the solar system.

For instance, bytownite, augite, the Bruderheim meteorite, enstatite, troilite, and pyrrhotite are possibilities. Bytownite and augite represent terrestrial analogs of the major constituents of the lunar surface; the Bruderheim meteorite is the most commonly occurring on the earth; enstatite, troilite, and pyrrhotite have been postulated as primary condensation products in the early stages in the formation of the solar system. The optical complex indices of refraction are listed in Chapter V, and Mie scattering calculations of extinction may be made. For simplicity, the particles are assumed to be spheres, although, in nature, this is an approximation. A relatively strong absorption is necessary in the candidate materials at 0.22 μm to produce the observed peaking shown in Fig. IX-1. Only bytownite, the high iron–nickel Bruderheim meteorite, and enstatite then remain as candidates.

The radius of the interstellar medium particles must be much less than

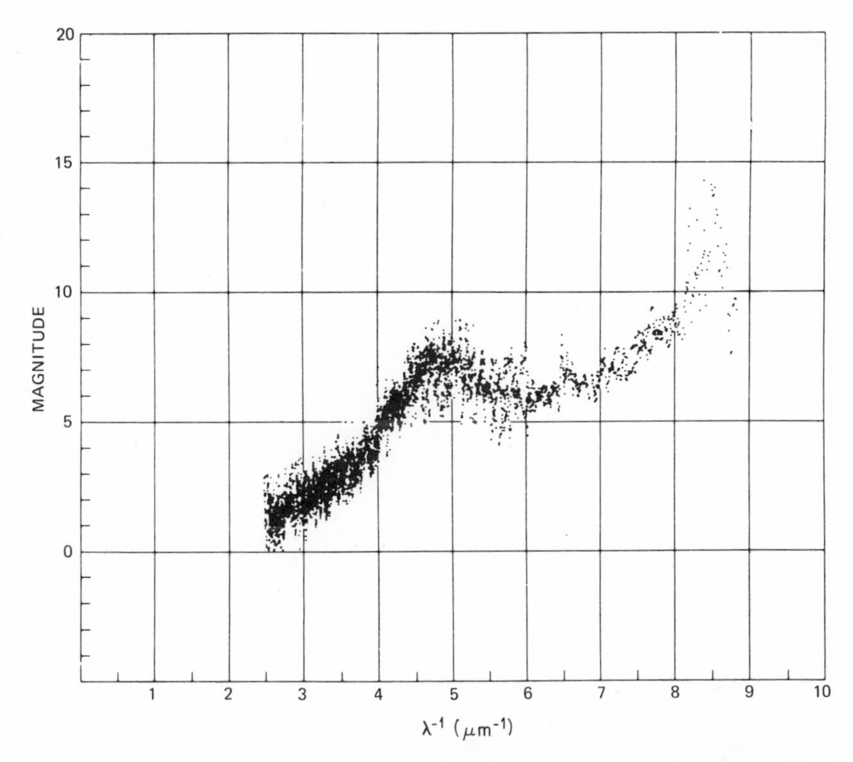

FIG. IX-1 Interstellar extinction in magnitudes as a function of inverse wavelength determined from the stars ζ and ε in the constellation Perseus. Curve is normalized to B − V magnitude = 1 at V ≅ 0. (Stecher, *The Astrophysical Journal*, **157**, L125, 1969. Copyrighted 1969 by The University of Chicago Press. All rights reserved.)

the incident radiation to produce the absorption feature at 0.22-μm wavelength and the strong extinction decrease with wavelength that are observed. One possibility appears to be bytownite (Egan and Hilgeman, 1975b), for a combination of particle radii between 0.1 and 0.0001 μm (see Fig. IX-2). A combination of this range of particle sizes will produce a median curve closely matching that observed. The solid lines in Fig. IX-2 are boundaries of the range of variation in interstellar extinction shown in Fig. IX-1. Further, a bytownite particle radius of 0.003 μm fulfills the additional requirement that the albedo be ≤ 0.2 at 0.22-μm wavelength (Witt and Lillie, 1973).

The double peak shown in Fig. IX-2 does not appear in the observations, but this could be the result of a distribution of nonspherical particles shifting the double peaks variously, and thus smearing the two peaks into one. Greenberg (1972) has shown this wavelength shift for oblate and prolate spheroids. This would be in agreement with the current thinking that

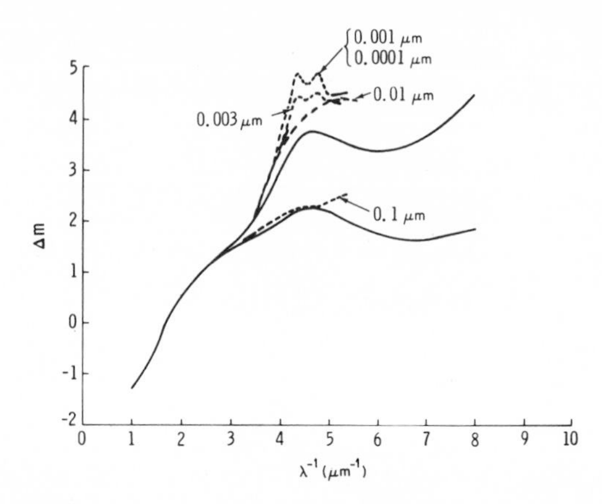

FIG. IX-2 Differential absorption versus wavelength for an interstellar medium consisting of various particle sizes of bytownite. [From Egan and Hilgeman (1975b).]

suggests the existence of particle sizes on the order of 0.005 μm (Greenberg and Hong, 1974).

This discussion is not meant to be the final episode in the saga of the interstellar medium. Recently, organic molecules have been suggested as possible constituents of the interstellar medium; in particular, polysaccharides appear capable of matching the 3- and 10-μm bands but not the ultraviolet (Hoyle and Wickramasinghe, 1977; Hoyle *et al.* 1978; Egan and Hilgeman, 1978b).

We feel that there is still much room for further clarification in terms of interstellar medium models and further interstellar extinction observations. It should be borne in mind that the data we have presented cover a narrow wavelength range for only several materials. However, the procedures presented may be applied to more candidate materials over a wide wavelength range to ultimately aid in the identification of the composition of the interstellar medium.

NOTES ON SUPPLEMENTARY READING MATERIAL

Allen (1963): Presents data on interstellar clouds and interstellar grains (pp. 251 ff).

McLaughlin (1961): A good introduction to astronomy; Chapter 19 deals with interstellar matter, the origins and composition.

Wickramasinghe (1967): Presents an account of theories and observations on the nature of the interstellar medium, with particular emphasis on grain theories.

CHAPTER X □ THERMAL ENERGY COLLECTORS: COATING CHARACTERISTICS

A. INTRODUCTION

Thermal energy collector design is in large part the application of the optics of inhomogeneous materials. The efficiency and visual and thermal properties of the collector surfaces depend strongly on the optical complex indices of refraction and the surface microstructure.

The radiant energy of the sun extending from the gamma ray region, through the x-ray, ultraviolet, visual, infrared, microwave, and radio regions, ultimately constitutes a source of heat. Most of the solar energy is in the visual range (~ 0.5 μm), and it may be collected by visually black surfaces; to prevent the reradiation of the infrared energy back into space by the heated black surfaces, either a selective filter can be placed above them or the "black" coating itself may be made to radiate very little infrared. A combination of these two approaches may be used plus the addition of optical-energy concentration devices such as reflectors or lenses.

The importance of solar energy is increasing as the cost of fossil fuel continues to rise. The earth's atmosphere is, for all practical purposes, transparent between 0.3- and 2.0-μm wavelength. At the temperature of boiling water (100°C), the peak blackbody emission wavelength λ_{max} is 7.8 μm from Wien's law:

$$\lambda_{max} = 2898/T \qquad \text{(X-1)}$$

where T is the temperature of the emitter in degrees Kelvin.

Total radiation from such a surface, Q, is given by the Stephan–Boltzmann equation:

$$Q = \varepsilon\sigma T^4 \qquad \text{(X-2)}$$

where σ is the Stephan–Boltzmann constant, T the surface temperature in degrees Kelvin, and ε the thermal emissivity or emittance of the surface (≤ 1).*

The emissivity ε is related to the surface reflectance R in opaque materials by

$$\varepsilon = 1 - R \tag{X-3}$$

This relation is true for each wavelength considered, and is a representation of Kirchhoff's law: a good emitter ($\varepsilon = 1$) is a good absorber ($R = 0$). At any wavelength, the monochromatic emittance ε_λ is equal to the monochromatic absorption α_λ.

Thus, an energy selective surface, suitable for collecting solar radiation, would have a large α_λ in the visual range α_v and a small ε_λ in the infrared ε_i (i.e., both the "α_v/ε_i ratio" and α_λ should be large).

B. SIMPLE FILMS

Many surfaces have been described that have more or less fulfilled the requirements of a large α_v/ε_i; Tabor and co-Workers (Tabor 1951, 1955, 1959, 1962, 1966; Tabor *et al.*, 1961, 1964), have enumerated many, and, in particular, the "Tabor" coatings of nickel black. The properties of these surfaces obviously depend upon their optical complex indices of refraction as well as their surface structure, but basically they absorb strongly in the visual region and reflect strongly in the long wavelength infrared (Table X-1). The use of reflectance to characterize a surface is incomplete because surface geometry (which may drastically change the reflectance) is unspecified. However, the optical complex indices of refraction are not yet available for the many energy collector materials that exist at present. Therefore, the approach followed in this chapter is to limit the discussion of reflectance (scattering) properties of these surfaces to one based on published data and measured reflectance. The more-detailed calculations treating the geometry separately must await the measurements of appropriate complex indices.

Gold has a high α_v/ε_i ($=10.0$) but a low absorptance α_v ($=0.3$), as shown in Table X-1. It thus emits a small fraction of the energy absorbed, but preferred solar panels should have an α_v value between 0.9 and 1.0. This is required to lower the cost per square meter. It is important to recognize the difference between optical efficiency and cost effectiveness. Commercial

* The quantity σ has previously been used to denote the quantity of radiation scattered per unit distance and is unrelated to the usage in this chapter.

TABLE X-1 Thermal Properties of Selected Materials

Material	α_v	Temperature (°K)	ε_i	α_v/ε_i
Tabor solar collector chemical treatment of galvanized iron	0.885	309	0.122	7.25
Tabor solar collector chemical treatment 110-30 on nickel-plated copper	0.853	309	0.049	17.4
Gold plate on 7075 aluminum	0.3	278	0.03	10.0
Alcoa Brown (unexposed to sunlight)	0.85	293	0.3	2.8
Alcoa Brown (exposed to sunlight)	0.84	293	0.3	2.8
Alcoa Black (unexposed to sunlight)	0.93	293	0.333	2.8
Alcoa Black (exposed to sunlight)	0.89	293	0.3	3.0
Black enamel (Ferro Corp.)	0.88	293	0.80	1.1

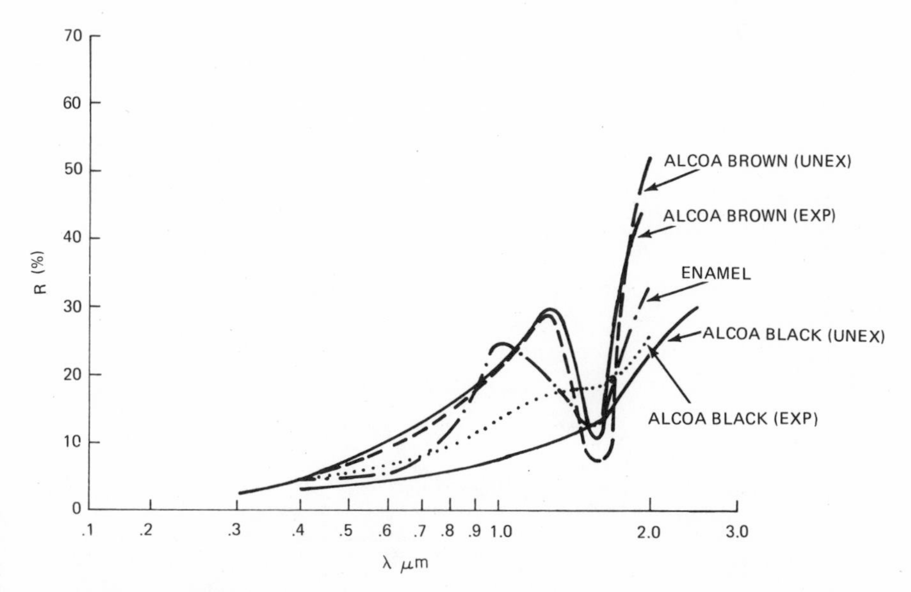

FIG. X-1 Total diffuse reflectance of selected thermal energy collector coatings between 0.3 and 2.7 μm.

collector design stresses cost per watt rather than watts per square meter. A relatively large infrared emittance can be counteracted by placing a selective filter above the solar energy collector panels that transmits the incident solar radiation and reflects the emitted infrared radiation; this increases the effectiveness of the solar energy collector panels even though they have a high emittance.

Other solar collector coatings listed in Table X-1 are Alcoa Brown, Alcoa Black, and a black enamel. These materials have a high α_v/ε_i, and a reflectance between 0.3 and 2.7 μm as shown in Fig. X-1 (i.e., they are all dark in the visual range with an increased reflectance into the infrared region). The increased

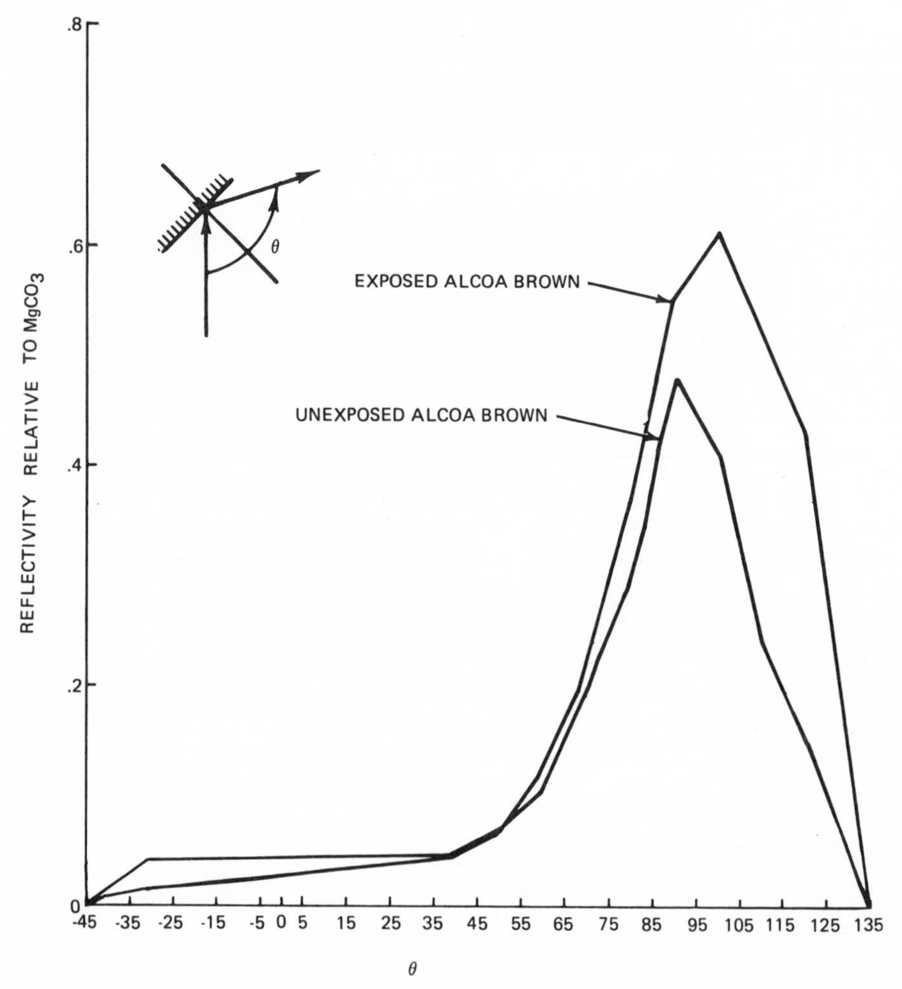

FIG. X-2 Reflectivity relative to $MgCO_3$ of selected energy collector coatings as a function of angle θ (shown in inset); incident angle, 45°, wavelength, 0.55 μm.

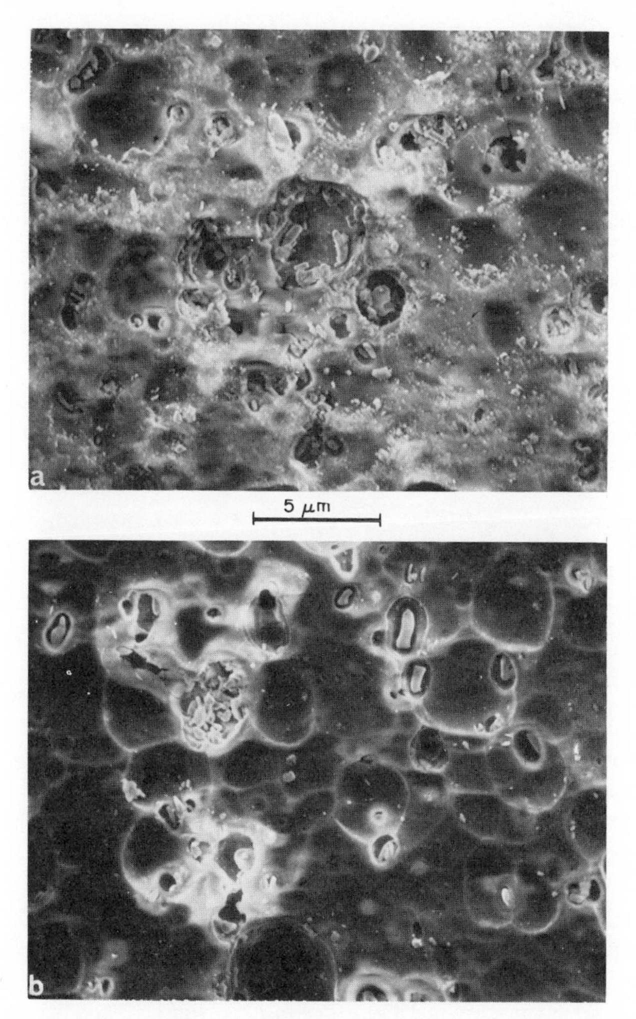

FIG. X-3 Scanning electron microscope photographs of Alcoa Black thermal energy collector coating (a) unexposed and (b) exposed.

infrared reflectance does not have much effect on the value of α_v because most of the sun's energy is in the visual range. The dip at about 1.8 μm is the result of water in the coatings, causing a low reflectance. The interesting fact shown by Fig. X-1 is that coatings may degrade with prolonged exposure (about 1 year) to sunlight. Thus, the Alcoa coatings became generally more reflecting and had lower α_v values (Table X-1) following a prolonged exposure to sunlight. In addition, the angular reflectance characteristics of the surfaces changed. Figure X-2 reveals that, for example, one change is an increase in the near specular reflection of an Alcoa Brown surface. The cause is surface degradation as is revealed by a scanning electron microscope photograph of an Alcoa Black surface before and after exposure (Fig. X-3). One observes that the surface roughness is diminished by the solar radiation, perhaps as the result of high concentrations of thermal radiation that have not been conducted away by the substrate. This serves to point out that caution must be taken in the application of these various collector surfaces. These thermal radiation concentrations could have fused the micrometer-sized particulates into the surface, decreasing the surface roughness, increasing the reflectivity, and decreasing the α_v.

The black enamel should be unaffected by exposure because the surface is already specular.

C. GRADIENT REFRACTIVE INDEX FILMS

In order to obtain a large α_v, the reflectance must be kept low in the energy-absorbing spectral region between 0.35 and 2.0 μm. If the surface reflectivity of a solar collector could be kept $<\frac{1}{2}\%$ by "matching" the surface impedance to that of the incident medium, the objective of $\alpha_v \cong 1.0$ could be obtained. This matching could be accomplished by a gradient refractive index antireflection film such as is produced by a chemical etch/leach process on glasses (Minot, 1976). The broadband, antireflecting film is a single porous layer primarily composed of silica. Typical reflectance curves for films of various structures are shown in Fig. X-4 as a function of d/λ (i.e., the thickness–wavelength ratio). The parameter Δn_2 is the change in the index of refraction within the transition film of thickness d. Cases A and B are the result of discontinuities in the refractive index at both surfaces of the gradient film. The discontinuity between the film and substrate has been made zero for case C; for case D, there are no discontinuities at either film surface.

If the films can be made economically in production, they could result in high α_v, and a trend toward a high ε_i could be counteracted by a filter placed over the collector.

D. COMPOSITE MATERIAL FILMS

Another optical system that employs a spatially varying dielectric constant of scale much smaller than the wavelength of the incident radiation has been described by Abeles and Gittleman (1976). This differs from the gradient refractive index in that granular metals such as silver, silicon, and germanium are introduced into an insulator such as silicon dioxide, silicon

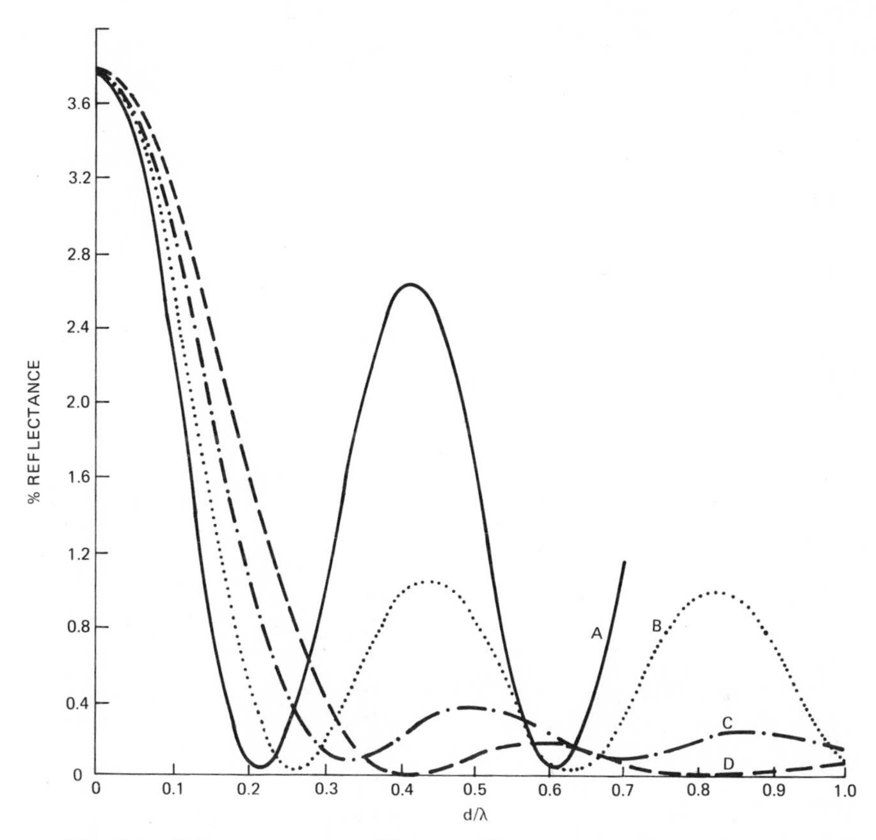

FIG. X-4 Reflectance versus d/λ for gradient refractive index films where the refractive index gradient Δn_2 is made progressively larger from curve A through curve D. [From Minot (1976).]

carbide, or aluminum oxide to produce a labyrinth structure. The mathematical representation of the interaction of incident electromagnetic radiation with such a structure is forbidding, and approximate solutions have been used; for a 640-Å-thick Ag–SiO_2 film, the optical density, i.e., $\log_{10}(I_{out}/I_{in})$, is represented better by the Maxwell-Garnett theory, (Maxwell-Garnett, 1904, 1906) than the effective medium theory. The Maxwell-Garnett theory assumes that spherical grains (here Ag) are embedded in an insulating matrix (here SiO_2) (see Fig. X-5). As an example of a material for a solar collector, Ge embedded in an Al_2O_3 matrix has a $\log_{10}(I_{out}/I_{in})$, is represented better by the Maxwell-Garnett theory low reflectance (~ 0.2) in the visual range and high reflectance (~ 0.9) in the infrared, unlike pure germanium. With a suitable selection of a composite system, α_v may be made high.

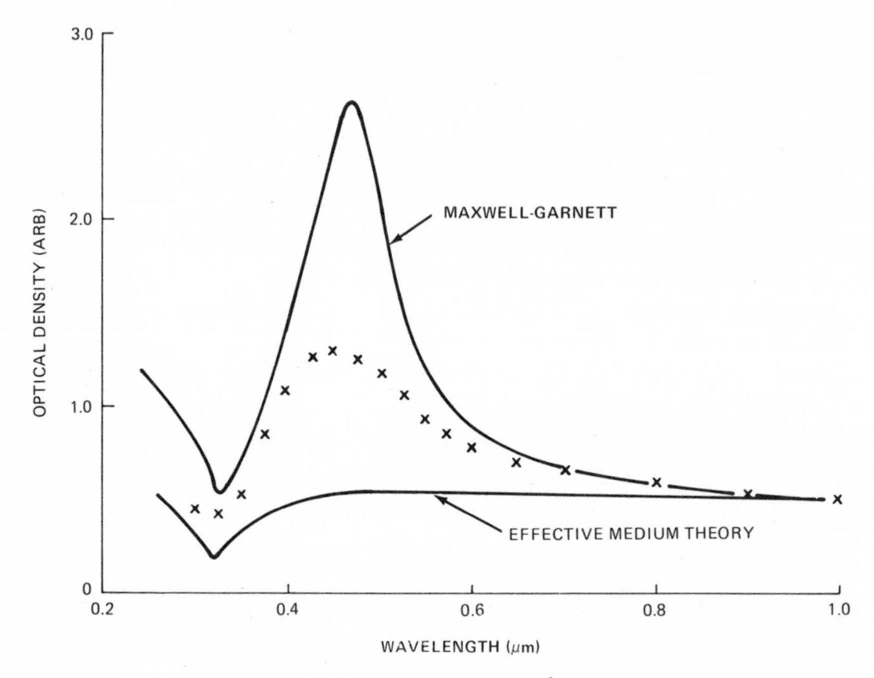

FIG. X-5 Optical density of Ag–SiO_2 film (640-Å thick) containing 0.39 volume fraction Ag. (——) Theoretical; (×) experimental points. [From Abeles and Gittleman (1976).]

E. ENHANCEMENT TECHNIQUES

A number of techniques have been suggested which serve to augment a collector's optical efficiency. These can be used to correct for the deficiencies of an inexpensive coating or enhance the overall efficiency of an already suitable coating. For instance, a relatively large infrared emittance can be counteracted by placing a selective filter above the solar energy collector panels that transmits the incident solar radiation and reflects the emitted infrared radiation.

Another technique involves careful control of the coating thickness to approximately one wavelength. This has the effect of allowing a coating that is normally black at all wavelengths to absorb in the visual where the wavelength is short, but be transparent (low emitting) in the infrared, where the wavelength is longer than the surface thickness.

NOTES ON SUPPLEMENTARY READING MATERIAL

Allen (1963): Reference material on black-body radiation, tables of the Planck functions, and the characteristics of solar radiation.

Born and Wolf (1959): Discussion of dielectric and metallic films, optical properties of dielectrics and metals, absorption and reflection of radiation.

Kruse *et al.* (1962): An integrated presentation on the nature of infrared radiation, such as its reflection, refraction, absorption, diffraction, and scattering by various media; also presents information on infrared components, such as windows, detectors, and sources.

Richtmyer *et al.* (1955): Discussion of quantum theory, and black-body radiation and the origin of spectral lines.

Wolfe (1965): A presentation of infrared techniques, mainly developed during World War II, as applied to military problems, in particular; a wealth of data is presented in tabular and graphical form.

CHAPTER XI □ COATINGS AND PAINTS

A. INTRODUCTION

We are surrounded by surface coatings, man-made and natural. The development of coatings and paints is ruled by the empirical approach, with a certain amount of "black art." Fabrication would be simplified if one could quantitatively calculate the angular as well as color reflectance properties of mixtures from a knowledge of the fundamental optical properties of the constituents. Paint- and coating-related industries would greatly benefit if a theoretical approach could be developed. The effect on reflectance of the binder, the fineness of grind of the pigments, the extenders, and the significance of two-component (epoxy) formulations remain unclear. Also, impurities have perhaps been attributed too significant a part in yielding a certain reflectance or emittance.

In this chapter, we will discuss, initially, a purely heuristic representation of the reflectance of compacted powder mixtures, such as would be representative of paints and coatings, without regard for the causes of absorption and scattering, in terms of the fundamental optical properties of the materials. Following this, a discussion will be made of some examples of the tailoring of the paint or coating reflectance to meet color or camouflage requirements, with particular emphasis on attaining low infrared reflectance. The tailoring depends upon the availability of pigments having as nearly as possible the desired absorbing properties as a function of wavelength. The amount of scattering that is introduced will affect the overall brightness and flatness (i.e., nonspecularity) of the surface. Additives may be employed to influence the color, such as dyes, colored glass beads, or semiconductors. The diffuseness (or flatness) of a coating is primarily influenced by the structure of the optical surface, which is taken to include effects extending below the actual physical surface of the coating.

B. COMPACTED POWDER MIXTURES

The reflectance of binary powder mixtures may be divided into three classes by characterizing the opacity of each of the two phases. They are mixtures of particles that are (1) opaque–opaque, (2) opaque–transparent, (3) transparent–transparent.

For a surface consisting of opaque particles, the reflectance primarily depends upon the particle reflectance, multiple scattering being small. Thus, for a surface consisting of a mixture of two kinds of opaque particles, the reflectance will depend upon the relative surface areas of the two components. Following this line of thought, an analysis by Schatz (1967) predicts the reflectance of an opaque–opaque particle mixture by calculating the effective surface area of component B in a mixture of two components A and B:

$$S_B = W_B \rho_A d_A/(W_B \rho_A d_A + W_A \rho_B d_B) \qquad \text{(XI-1)}$$

where S is the percent surface area, W the percent by weight, ρ the average particle density, and d the average particle diameter.

At a specific wavelength, the resulting reflectance R_m of a binary mixture is given by

$$R_m = S_A R_A + S_B R_B \qquad \text{(XI-2)}$$

For opaque–transparent mixtures, the analysis is not as simple because the transparent particle scatters radiation. Then, when R_A is the transparent component,

$$R_m = S_A{}^2 R_A + S_A S_B R_A' + S_B R_B \qquad \text{(XI-3)}$$

where

$$R_A' = (n_A - 1)^2/(n_A + 1)^2 \qquad \text{(XI-4)}$$

and n_A is the refractive index of material A.

The quantity R_A' represents the average reflectance of a particle of material A surrounded by particles of material B.

For mixtures of transparent particles, the reflectance R_m becomes

$$R_m = S_A R_A + S_B R_B + S_A S_B[R_A'(1 - R_B/R_A) + R_B'(1 - R_A/R_B)] \qquad \text{(XI-5)}$$

where R_B' is defined similarly to R_A'. These relations produce good agreement with experimental data in most cases.

C. TAILORING REFLECTANCE

The primary considerations for a paint or coating are the color (i.e., the wavelength dependence of reflectance) and the angular scattering properties (i.e., the diffuseness or glossiness).

The color is produced by the absorption and scattering properties of the surface arising from two sources, a surface reflection component and a body and surface scattering component (Wendlandt and Hecht, 1966). The surface reflection component is due to the difference in the complex index of refraction between the coating and the ambient medium. For most coatings, the refractive portion of the index has the dominant effect on the surface reflection. Also, since the refractive component of the index of many coatings is low (i.e., 1.5 to 2.0) the surface contribution to reflectance is low. It can be highly specular, however, if the surface roughness is of a scale smaller than that of the wavelength. To eliminate the specularity, the surface roughness can be made high. The other component, body and surface scattering, can be a dominant contributor to the reflectance. The magnitude of the reflectance due to body scattering depends on the ratio of the absorption coefficient to the scattering coefficient (sum of body and surface scattering) as shown in Fig. XI-1. This figure is obtained by using Eq. (II-19) for R_∞ and Eq. (II-18) for β, and substituting an appropriate range of values for the ratio of the absorption coefficient to the scattering. As seen from the figure, to obtain a low reflectance, a ratio of absorption to scattering coefficient of at least 1.0 would be required. This ratio can be achieved either by increasing the absorption coefficient, or by decreasing the scattering coefficient.

An optimum particle-size distribution and compaction may be found that will make the scattering lower in one spectral region and yet sufficiently high in another spectral region to give a desired coloring. For example, the scattering coefficient can be lowered at longer wavelengths ($\sim 1\ \mu$m) by

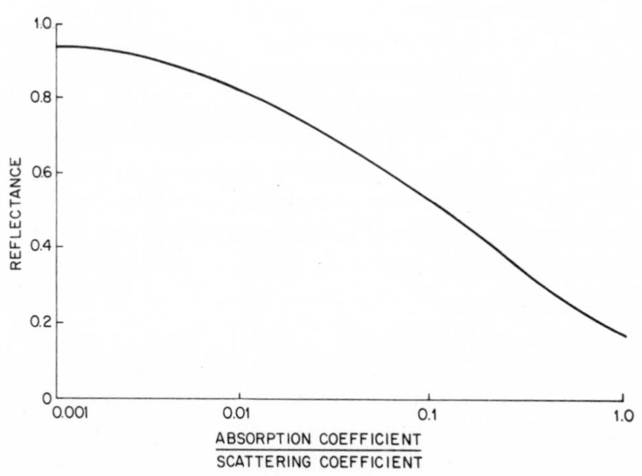

FIG. XI-1 Example of effect of ratio of absorption to scattering on reflectance.

selecting a particle size for the coating pigment that is larger than about 1 μm and/or by varying the pigment concentration.

Another means for increasing the absorption to scattering ratio is to use materials that have a high absorption in one region compared to that in another region. For instance, military applications often require a low infrared reflectance. Many materials that absorb well in the infrared are also nonspectral absorbers in the visible region of the spectrum and, therefore, have a gray or black color. There are exceptions to this, however. Some transition-metal ions have localized absorption bands in the visible and infrared which depend on their valence and crystal structure. Since these absorption bands tend to be narrow, combinations of absorbers would have to be used to achieve a desired color. By making some materials nonstoichiometric, the infrared absorption can also be increased. Reduced rutile (TiO_{2-x}), for example, has an absorption peak from polarons at about 1 μm that is not present in TiO_2. Another example is single crystal MoO_3 which has absorption peaks at 3930, 4130, and 4250 Å wavelengths dependent upon the polarization of the radiation (Deb, 1968). Ultraviolet irradiation of a thin film of this material produces a broad color-center band with a maximum at 8700 Å (Deb, 1968). Analogous effects occur with WO_3 (Deb, 1973).

Other semiconductors that represent potential red- and infrared-absorbing pigments are FeS_2, CoS_2, NiS_2, and CuS_2; their optical complex indices of refraction are presented in Fig. XI-2 (Bither *et al.*, 1968). Iron pyrite, FeS_2, has a golden hue, CuS_2 appears lead gray, and CoS_2 and NiS_2 are black. If a low infrared reflectance were desired at ~ 2 μm, then CoS_2, NiS_2, or CuS_2 would be suitable candidates for pigments. Naively, one might assume that the same absorbing effect could result from using powders of infrared-absorbing visual-transmitting glasses, such as the ones shown in Fig. XI-3, as pigments. In truth, the glasses do not have sufficient absorption to achieve sufficient hiding when used as a pigment, but act more as extenders. The same difficulty applies to various absorbing dyes (Fig. XI-4) used in lucite particles if the particles are intended as a pigment; however, the dyed lucite particles could serve as a yellowing retarder to attenuate the scattered ultraviolet radiation reaching a white pigment.

The effect of surface roughness, on the other hand, is to produce scattering which affects the brightness (tint) of a hue. As surface roughness increases, with the scale of roughness less than the wavelength of the incident radiation, the surface will become brighter.

Low specularity is achieved by increasing the surface roughness on a scale larger than the wavelength. Techniques such as increasing pigment volume concentration and adding flatting agents have proven successful in the visual. Another possibility is that changes in the shape of the spatial frequency curve of the surface roughness [also known as the modulation transfer

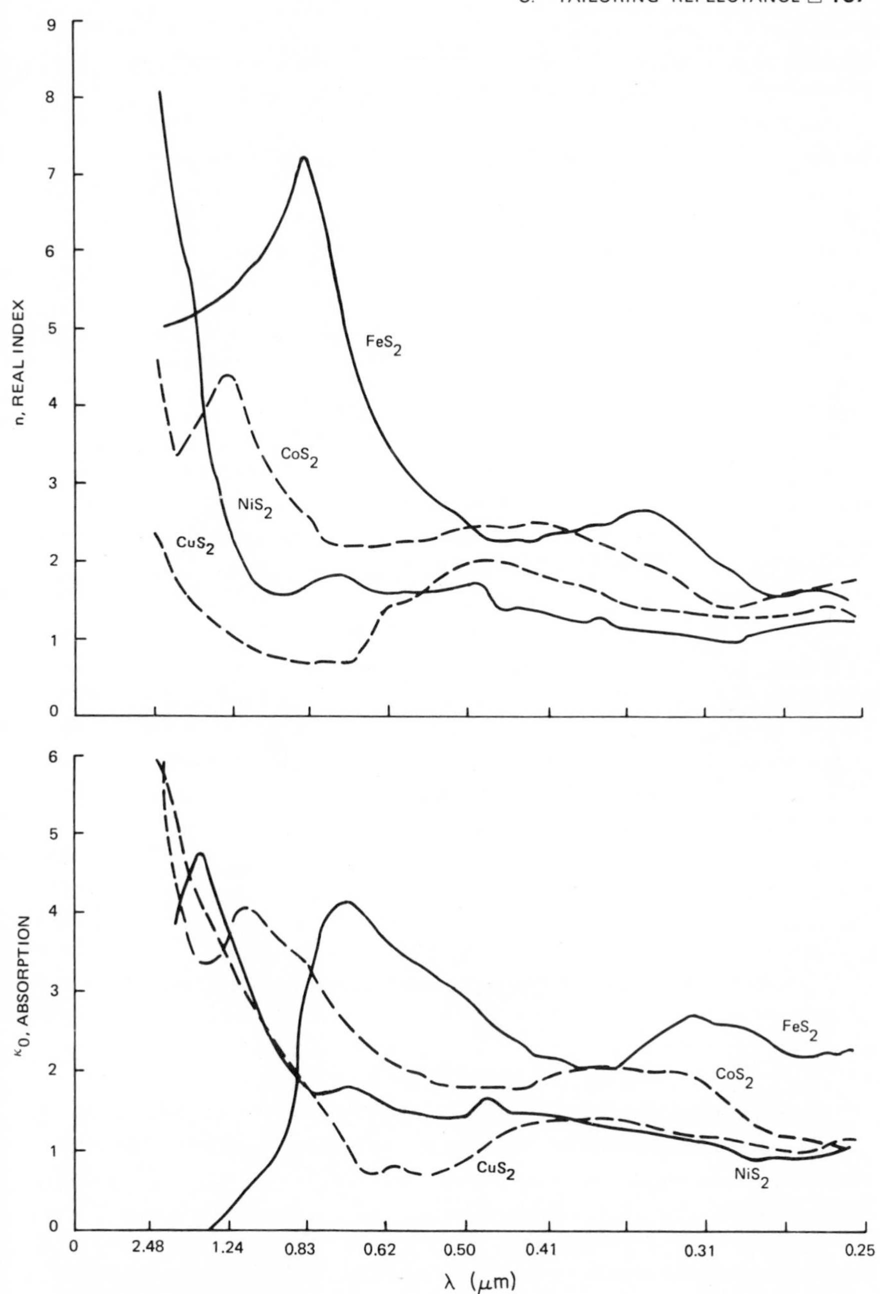

FIG. XI-2 Real and imaginary parts of the complex refractive index for FeS_2, CoS_2, NiS_2, and CuS_2. [From Bither *et al.*, *Inorganic Chemistry* **7**, 2208 (1968). Copyright by the American Chemical Society.]

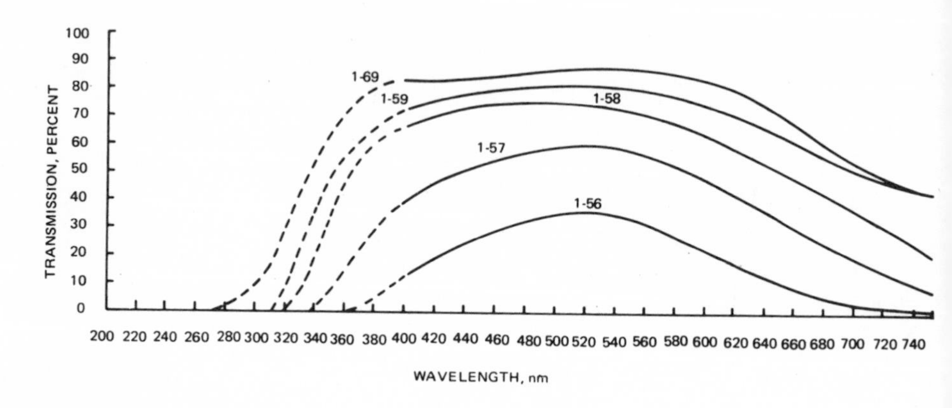

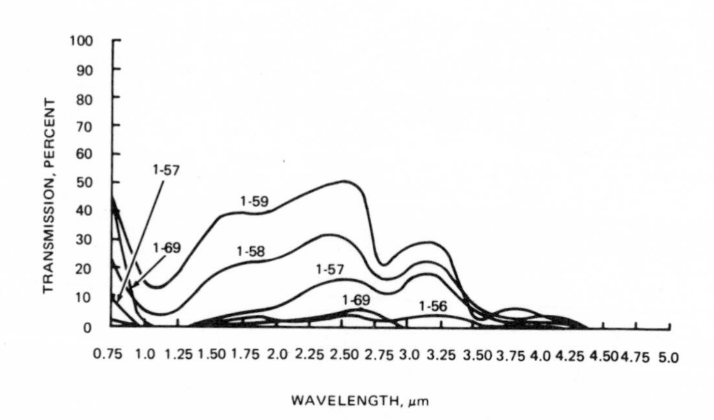

FIG. XI-3 Transmission of Corning infrared-absorbing visual-transmitting glasses. (From Corning Glass Color Filters Bulletin.)

function (MTF)] can affect specularity. For instance, increasing either the high or low spatial frequency amplitudes decreases specularity, but each in a different way. Increasing the high spatial frequencies produces a higher diffuse reflectance (Rayleigh scattering) with surface brightening, while increasing the low spatial frequencies will just spread the specular peak (quasi-specular scattering). We have described some general approaches and given some examples of tailoring the reflectance properties of coatings and paints. A much wider assortment of pigments is available beyond those discussed in this chapter.

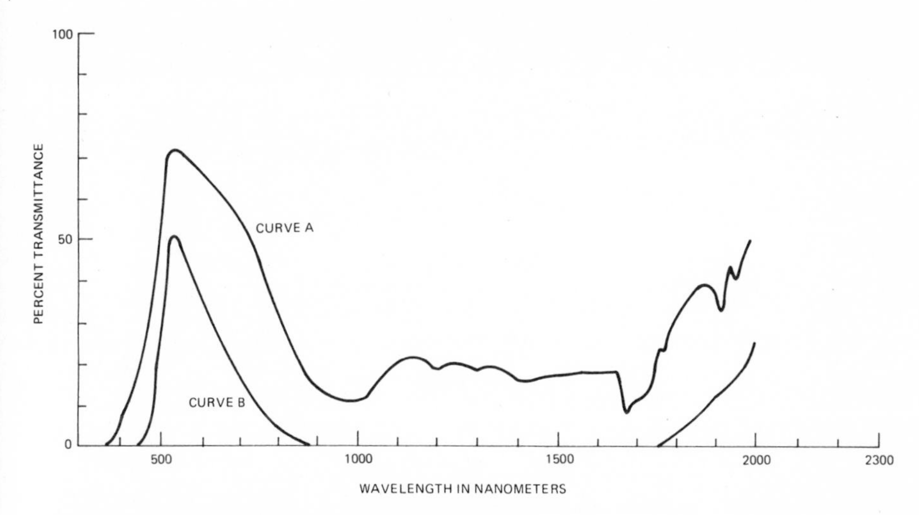

FIG. XI-4 Cyasorb IR-117 light absorber and UV-24 yellowing retarder in poly-(methyl methacrylate). Curve A, 0.03% IR-117 light absorber and 0.2% UV-24 yellowing retarder, plastic thickness 2 mm; curve B 0.1% IR-117 light absorbing, plastic thickness 1.8 mm. From American Cyanamid Company Bulletin No. 7-2701-3 M-2/77: Plastic Additives. (Reproduced by permission of American Cyanamid Co., Organic Chemicals Division, Bound Brook, New Jersey.)

NOTES ON SUPPLEMENTARY READING MATERIAL

Burns (1970): Theory and measurements of spectral reflectance of silicates containing transition metals; explains wavelength variations in reflectivity.

Seitz (1940): Presents theoretical and experimental aspects of the electronic structure of solid bodies that can affect the reflectance characteristics.

CHAPTER XII □ REMOTE MINERAL EXPLORATION

A. INTRODUCTION

A very desirable accomplishment would be the location of minerals by optical remote sensing, armed with the optical complex indices of refraction of all possible minerals. This ideal has many things going against it, in particular the effects of erosion, weathering, ground cover by foliage or water, and the fact that many minerals of interest are beneath the surface. Often geologists are forced to rely on secondary, interpretative techniques such as inferences from outcroppings, faults, and other geological features. An understanding of the basic properties of how the rocks have been formed has also been of significance in mineral exploration; for instance, the new theories of plate tectonics have led to the discovery of minerals.

An important application of optical remote sensing is to mineral exploration in space, where, excluding landers, all the surface information comes in the form of electromagnetic radiation, much of which is in the wavelength region between the ultraviolet and infrared. Very high spectral and spatial resolution optical instruments exist at these wavelengths, making data acquisition of secondary concern. The real problem lies in the interpretation of the data.

Nature confounds the recognition problem with space erosion from meteorite bombardment of surfaces by, for instance, producing a regolith that may not be representative of the primordial surface. Also, surface minerals are rarely of a single substance. The minerals may be analogous neither to terrestrial ones nor to perfect single crystals.

Remote mineral exploration consists of several phases, starting with the reconnaissance phase, in which, perhaps, regional mapping, airborne or orbiter geophysics and geochemistry, and geological interpretation are performed. Having thus selected target areas, detailed mapping and ground truth-related activities would then be initiated.

The lunar surface is the most extensively investigated extraterrestrial surface to date, and it will be discussed, using the results of ground-based telescopic observations with correlations to the optical properties of actual lunar samples in the wavelength range 0.4–2.5 μm. This will be followed by a brief discussion of Martian remote sensing using ground-based telescopes, the NASA Lear Jet observatory, and the NASA Convair 990 aircraft.

B. LUNAR REMOTE SENSING

The first detailed lunar mapping program was initiated in 1960 by the U.S. Geological Survey using earth-based telescopic observations. Such geographical maps of lunar areas can provide significant compositional information when combined with results of the lunar exploration programs. The general approach to lunar remote sensing has been phenomenological because neither the appropriate optical complex indices of refraction of lunar materials nor adequate surface models existed to permit detailed analyses to be made. Although adequate models are now available (Chapter II), the appropriate optical complex indices of refraction still have not been measured. However, implicit in present remote-sensing analyses is the effect of the optical complex indices of refraction (and surface structure) in causing reflectance variations. As a first approach, reflectance curve matching techniques to be described have enjoyed some limited success, and specialized instrumental techniques have been developed. The nonuniqueness of the curve matching has caused difficulties in achieving unambiguous identifications.

Mare, mare crater, upland, and bright upland crater spectral types have been defined by McCord *et al.* (1972), and these may be quantitatively related to the reflectance slope plot between 0.402 and 0.564 μm versus the intensity ratio between 0.564 and 0.948 μm (Charette *et al.*, 1977); a relationship was shown between the TiO_2 content of bulk lunar soils and the spectral curve slope between 0.402 and 0.564 μm. This is significant because basaltic units can be distinguished based on titanium content. A map of the lunar spatial distribution of the reflectance spectrum slope would give a map of the TiO_2 content and, thus, a map of the mare basic geologic units.

Lunar mapping has advanced with the development of a two-dimensional silicon-diode–vidicon imaging system with a spatial resolution on the moon of 2 km at the subearth point and with a photometric precision of better than 1%. This device has been used to obtain excellent images of the lunar surface at several wavelengths (McCord *et al.*, 1976) which have been used to produce maps of the TiO_2 content of the lunar soil. Johnson *et al.* (1977) have used the same silicon-diode vidicon to refine even further the lunar basalt classification and distribution using the 0.40–0.57-μm ratio, and to relate the 0.8–1.1-μm spectral features to agglutinate and iron content. As remarked previously, the optical complex indices of refraction are not employed in these analyses.

The first detailed optical complex index of refraction data on actual lunar samples was presented by Simpson and Bowie (1970) on ilmenite contained in them (Fig. XII-1). Unfortunately scattering calculations using these optical complex indices of refraction to represent the basalt reflectance would be incomplete without the use of the unspecified optical data on the other minerals in the basalt matrix from which the ilmenite was obtained. Results for two ilmenite samples are shown to illustrate the variability between specimens.

The technique of curve matching previously described has been criticized by Charette and Adams (1977); they point out that a given spectral reflectance curve topology may not be unique, while the individual absorption features

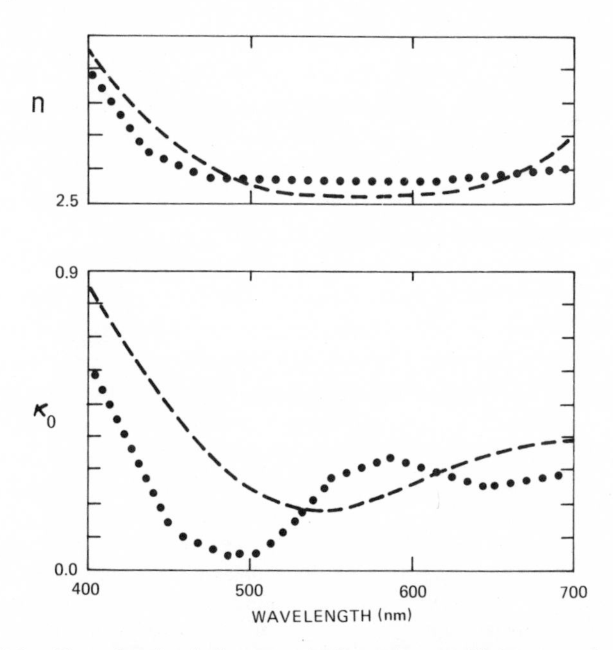

FIG. XII-1 The refractive index n and absorption coefficient κ_0 of two ilmenite samples in the visible spectrum for the ordinary vibration direction; ··· lunar sample No. 45.35.5; – – – lunar sample No. 85.4.16. [Reprinted with permission from Simpson and Bowie, *Science* **167**, 617 (1970). Copyright 1970 by the American Association for the Advancement of Science.]

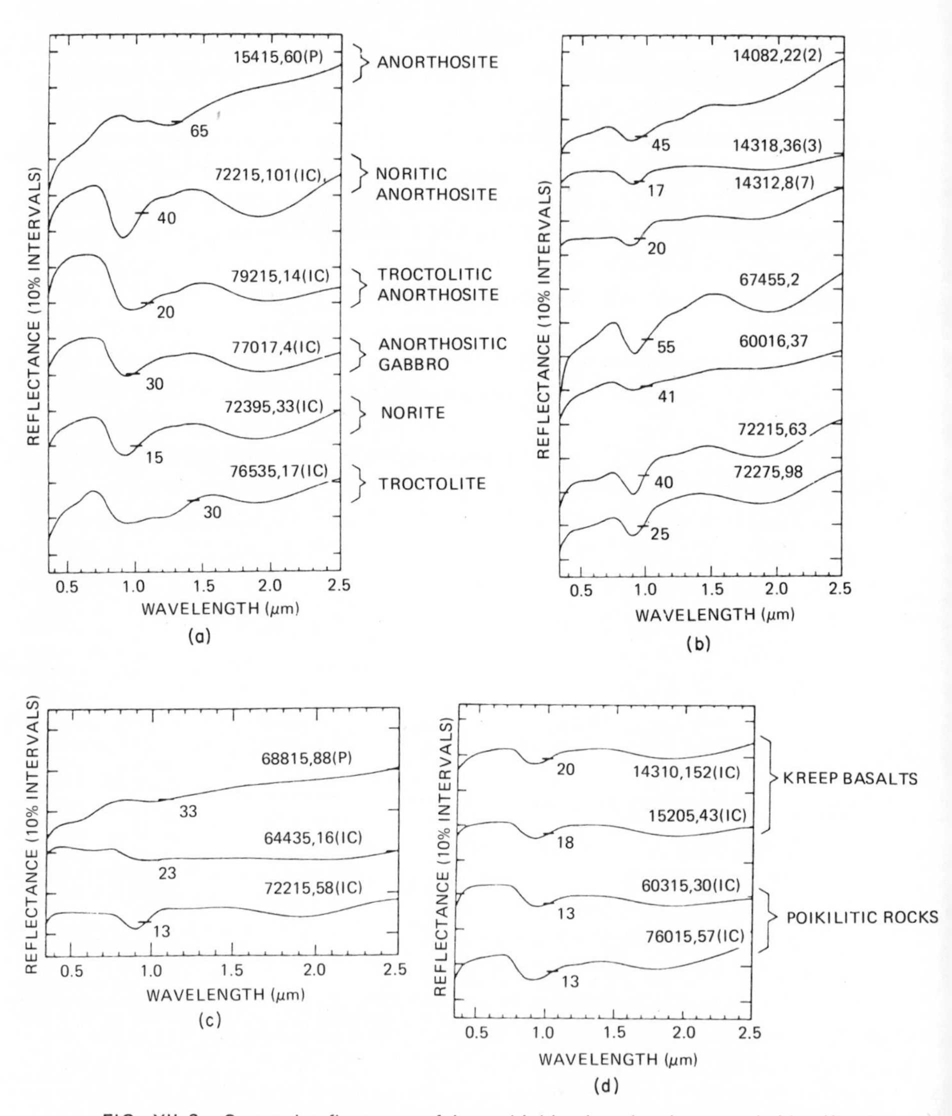

FIG. XII-2 Spectral reflectance of lunar highland rocks; lunar rock identification (as per Lunar Science Institute) shown on each curve as well as a reflectance calibration point for the sliding percentage scale; (a) coarse-grain rocks, (b) light-matrix breccia powders, (c) dark-matrix breccias, (d) nonmare melt breccias. Powders are designated by (P) and interior chips by (IC). A number of breccias are shown in order to indicate the considerable optical variability within this class. [Reprinted from Charette and Adams (1977), *Lunar Science VIII*, pp. 172–174. The Lunar Science Institute, Houston.]

in the curve resulting from specific mineral assemblages are unique. Egan and Hilgeman (1971) have also pointed this out, stating that even the individual absorption features may not be unique under varying surface geometries. Such mineral assemblages have absorption signatures in the visual and near-infrared range, and include pyroxene, plagioclase, olivine, ilmenite, and various colored glasses found in lunar soils and breccias. The distinguishing characteristics of lunar highland rocks are shown in Fig. XII-2 for coarse-grained rocks, light-matrix breccias, dark-matrix breccias, and melt rocks. For each of the four classes of samples (Fig. XII-2a–d), several samples are shown to illustrate the range of variability between samples. Thus, for instance, in the class of light-matrix breccias (Fig. XII-2b), there is a considerable range in the depth of the absorption band at 1 μm. This variation could be the result of variations in the powder particle geometry and/or the result of compositional variations. Although a number of authors have claimed that such spectra may be used to determine lithology, and attempts have been made to do this, the basic uncertainty of nonuniqueness remains with this approach. The use of models similar to those described in Chapter II will represent the next significant step in lunar remote sensing.

C. MARTIAN REMOTE SENSING

The surface of Mars has not been as thoroughly investigated as the lunar surface, and we still do not have any actual soil samples. The Viking landers furnished much detailed data on the soil composition, but, at present, the actual structure and chemistry is elusive. Among the possibilities that have been suggested for the composition of the Martian surface are limonite (Dollfus, 1961), montmorillonite (Hunt *et al.*, 1973), basalt (Adams and McCord, 1969), and andesite (Hanel *et al.*, 1972), with titanium dioxide as an additional constitutent (Pang and Ajello, 1976).

Earth-based observations of Mars have been made in the spectral region from 0.3 to 2.6 μm (200–400-km diameter regions) of bright areas. A typical spectrum in the region between 0.3 and 1.1 μm is shown in Fig. XII-3. There is an intense blue–UV absorption with just barely perceptible slope changes at $\sim$0.54 and 0.6 μm; such slope changes are characteristic of soils containing several percent ferric oxide. A weak absorption band is centered between 0.83 and 0.90 μm which is interpreted as due to pyroxene (Huguenin *et al.*, 1977).

Earth-based infrared spectra of the entire disk of Mars are shown in Fig. XII-4. The 0.83–0.90-μm band is identified as ferric oxide and the 0.90–0.96-μm band as pyroxene; the 1.22–1.25-μm band is attributed to

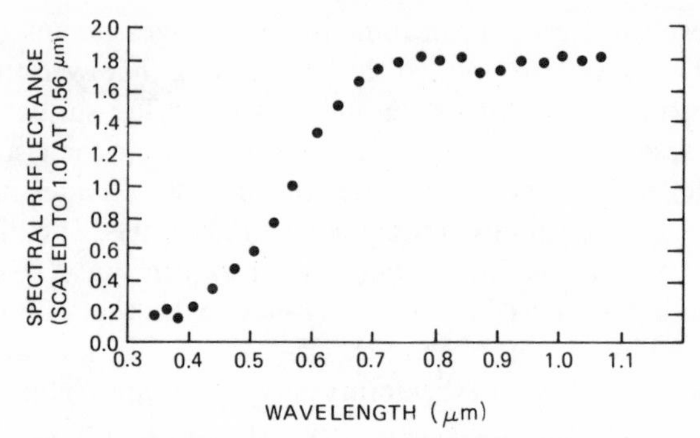

FIG. XII-3 Typical spectral reflectance of a Martian bright area. [Reprinted from Huguenin *et al.* (1977), *Lunar Science VIII*, pp. 478–480. The Lunar Science Institute, Houston.]

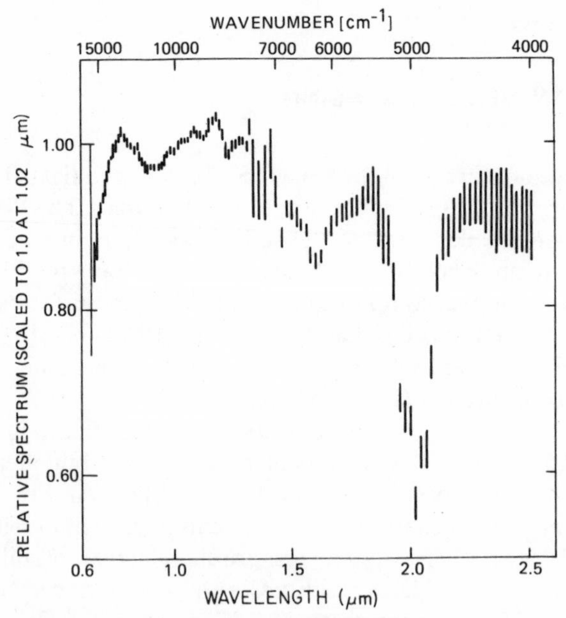

FIG. XII-4 Infrared spectrum of the integral disk of Mars. [Reprinted from Huguenin *et al.* (1977), *Lunar Science VIII*, pp. 478–480. The Lunar Science Institute, Houston,.]

feldspar (Huguenin *et al.*, 1977). Also evident is a broad H_2O ice band at ~1.6 and ~2.0 μm; superimposed are the narrower CO_2 bands in the Martian atmosphere at ~1.6 and 2.0 μm. The actual strengths of these CO_2 bands are confused by the nonlinearity of subtracting the CO_2 in the earth's atmosphere.

Medium-resolution spectral observations of Mars have been made from the NASA Convair aircraft using a 12-in. (30.5-cm) f/15 Cassegrain telescope and a 19-slot Hadamard mask in the exit plane of an Ebert–Fastie spectrometer (Houck *et al.*, 1973). The 2.0–4.0 μm spectrum was scanned in 0.112-μm steps as the aircraft flew between 38,000 and 41,000 ft (11.6–12.5 km). The precipitable water vapor was always less than 15 μm. These observations showed a broad absorption band in the spectrum of Mars, from which Houck *et al.* (1973) deduced bound water to be in the Martian surface material.

Subsequently, Egan *et al.* (1978) reported on similar high-altitude, higher-resolution observations (~17 cm^{-1}) of Mars using the NASA Lear Jet Astronomical Observatory; they found a transmission region between two CO_2 bands, centered at 3660 cm^{-1}, in which the depth of the Martian surface water of hydration band could be accurately defined. Then, by using the optical complex indices of refraction of appropriate terrestrial analogs and an appropriate surface model, the composition and structure of the Martian surface was inferred. They concluded that the optical property of the Martian surface layer material was consistent with a mixture of a basalt and a clay mineral—montmorillonite—in their wavelength region. The surface model was the six-flux Monte Carlo model discussed in Chapter VI.

Although the application of modeling techniques to remote mineral exploration is in its infancy, the pitfalls in the alternative phenomenological approach make this the logical direction for future investigations. Future applications of this approach can have great value in anticipated space mineral exploration.

NOTES ON SUPPLEMENTARY READING MATERIAL

Beckmann and Spizzechino (1963): A monograph on the scattering of electromagnetic waves from rough surfaces, such as irregular terrain, the sea, moon, and planets; an emphasis on radio waves.

Burns (1970): The reflectivity characteristics of silicates containing transition metals reveal characteristic properties that may be used to characterize minerals.

APPENDIXES

A. PROGRAM FOR CALCULATION OF OPTICAL CONSTANTS

IBM 360 FORTRAN IV program for calculation of optical complex indices of refraction using the modified Kubelka–Munk theory knowing total diffuse reflectance and transmittance and the real index of refraction (Egan *et al.*, 1973a).

```
C     MKM PROGRAM
      DIMENSION X(3,4),F(3,4),XX(2),AA(2),A(3,3),R(3),VAL(4)
      COMMON /EQUN/EN,DE,TACT,RACT
      COMMON /REF/VVV(2)
      DIMENSION HOL(10)
      DATA PI/3.141593/
      DATA EPS/5.E-7/,EPS1/1.E-8/,NIT/50/
      EXTERNAL EQU,NORM
      CALL ERRSET(207,256,2,2, 1,301)
      FP=4.0*PI
  100 READ(5,99,END=30)(HOL(I),I=1,10),AM,TACT,RACT,EN,DE,SB,ST,
     1AA(1),AA(2)
   99 FORMAT(10A1,5E10.0,4E5.0)
      WRITE(6,89)(HOL(I),I=1,10),AM
   89 FORMAT(1H1,24X,10A1,5X,'LAMBDA =',1PE17.5//)
      WRITE(6,98)SB,ST,EN,DE,AA(1),AA(2),TACT,RACT
   98 FORMAT(' BETA, TAU, N, D, BOUNDS, T ACTUAL, R ACTUAL'/
     11P8E16.5//)
      XX(1)=SB
      XX(2)=ST
      ND = 2
      NVD= 3
      IND= 0
      NT = NIT
      CALL REFALA(EQU,NORM,X,F,XX,AA,ND,A,R,EPS,EPS1,NVD,IND,NT,VAL)
      IF(IND)110,140,130
  110 WRITE(6,97)NT
   97 FORMAT('0NORMS INCREASED DURING ITERATION',I5//)
  115 IF(NT .LE. 1)GO TO 100
  120 DO 125 I=1,2
  125 XX(I)=VVV(I)
      NN = NT - 1
      WRITE(6,96)NN
   96 FORMAT(' RESULTS OF ITERATION',I5//)
      GO TO 160
  130 WRITE(6,95)NT
   95 FORMAT('0SINGULAR LINEAR SYSTEM DURING ITERATION',I5//)
      GO TO 115
  140 IF(NT .LT. NIT)GO TO 150
      WRITE(6,94)NT
   94 FORMAT('0THE MAXIMUM ITERATION',I5,' HAS BEEN FINISHED'//)
      NT = NT + 1
      GO TO 120
  150 WRITE(6,93)NT
   93 FORMAT('0CONVERGENCE AFTER',I5,' ITERATIONS'//)
  160 CALL TR(XX(1),XX(2),EN,DE,AA(1),AA(2))
      AL= (XX(1)*XX(2))/(2.*DE)
      ES=AL/(XX(1)**2)-AL
      WRITE(6,92)
   92 FORMAT(' FINAL ALPHA, S, BETA, TAU COMPUTED T, R  ACTUAL T, R'//)
      WRITE(6,91)AL,ES,XX(1),XX(2),AA(1),AA(2),TACT,RACT
   91 FORMAT(1P8E16.5///)
      ENK = (AL*AM)/FP
      WRITE(7,88)(HOL(I),I=1,10),AM,TACT,RACT,AL,ES,ENK,EN
   88 FORMAT(10A1,1P7E10.3)
      WRITE(6,87)ENK
   87 FORMAT(' EN KAPPA =',1PE16.5)
      GO TO 100
   30 STOP
```

```
      END
      SUBROUTINE NORM(V,VN,NDI)
      REAL * 4 V(1)
      VN = SQRT(V(1)**2 + V(2)**2)
      RETURN
      END
      SUBROUTINE EQU(X,Y)
      REAL * 4 X(1),Y(1)
      COMMON /EQUN/EN,DE,TACT,RACT
      CALL TR(X(1),X(2),EN,DE,T,R)
      Y(1) = T - TACT
      Y(2) = R - RACT
      WRITE(6,9)T,R,TACT,RACT,Y(1),Y(2)
    9 FORMAT('0IN EQU  T,  R,  TACT,  RACT,  Y(1),  Y(2)'/1P5E16.5//)
      RETURN
      END
      SUBROUTINE TR(SB,ST,EN,DE,T,R)
      REAL * 8 BET,RIN,TAU,GAM,RO,RI,RE,TC,RC,TD,RD,TT,RR,
     1ALD,ESD,END,DED
      BET=SB
      TAU=ST
      END = EN
      DED = DE
      ALD= (BET*TAU)/(2.D0*DED)
      ESD=ALD/(BET**2)-ALD
      AL =ALD
      ES=ESD
      RIN = (1.D0 - BET)/(1.D0 + BET)
      GAM=(ALD+ESD)*DED
      RO = ((END - 1.D0)/(END + 1.D0))**2
      RE = -.4399D0 + .7099D0 * END - .3319 * END** 2 + .0636D0*END**3
      RI = 1.D0 - (1.D0 - RE)/END**2
      TC =((1.D0 - BET**2)/(8.D0*BET**2 - 2.D0))*
     1(((RIN - 3.D0*RIN**2)* DEXP(-(2.D0*TAU + GAM))+
     2(3.D0 - RIN)*DEXP(-GAM)+
     33.D0*(RIN**2 -1.D0)*DEXP(-TAU))/
     4(1.D0 - RIN**2 * DEXP(-2.D0*TAU)))
      RC =((1.D0 - BET**2)/(8.D0*BET**2 - 2.D0))*
     1(((RIN**2 - 1.D0)*DEXP(-(TAU+GAM)) +
     2DEXP(-2.D0*TAU)*(3.D0*RIN - RIN**2) + 1.D0 - 3.D0*RIN)/
     3(1.D0 - RIN**2 * DEXP(-2.D0*TAU)))
      TD =((1.D0 - RIN**2)*DEXP(-TAU))/
     1(1.D0 - RIN**2*DEXP(-2.D0*TAU))
      RD = RIN *((1.D0 -DEXP(-2.D0*TAU))/
     1(1.D0 - RIN**2*DEXP(-2.D0*TAU)))
      TT = (((TC + RO*RC*DEXP(-GAM)) *
     1(1.D0 - RI*RD) + RI*TD*(RC + RO*TC * DEXP(-GAM)))*
     2(1.D0 - RO)*(1.D0 - RI)/
     3((1.D0 - RO**2*DEXP(-2.D0*GAM))*
     4((1.D0 - RI*RD)**2 - RI**2 * TD**2))) +
     5(1.D0 - RO)**2 * DEXP(-GAM)/(1.D0-RO**2*DEXP(-2.D0*GAM))
      RR = ((RC+RO*TC*DEXP(-GAM))*(1.D0-RI*RD) +
     1RI*TC*TD + RI*RO*TD*RC*DEXP(-GAM)) *
     2(1.D0 - RI)*(1.D0-RO) /
     3((1.D0-RO**2*DEXP(-2.D0*GAM))*((1.D0-RI*RD)**2 - RI**2*TD**2)) +
     4((1.D0-RO)**2*RO*DEXP(-2.D0*GAM)/(1.D0-RO**2*DEXP( -2.D0*GAM)))+RO
      T = TT
      R = RR
      WRITE(6,9)
    9 FORMAT(' AL,ES,EN,DE,T,R'/' BET,RIN,TAU,GAM,RO,RI,RE'/
```

```
     1' TC,RC,TD,RD,TT,RR'//)
C     WRITE(6,8)AL,ES,EN,DE,T,R,BET,RIN,TAU,GAM,RO,RI,RE,
C    1TC,RC,TD,RD,TT,RR
    8 FORMAT( 1P6E17.5//1P7D17.5/1P6D17.5//)
      RETURN
      END
      SUBROUTINE REFALA (EQU,NORM,X,F,XX,AA,ND,A,R,EPS,EPS1,NVD,IND,NT
     *,VAL)
      COMMON /REF/VVV(4)
      DIMENSION X(NVD,1),F(NVD,1),XX(1),AA(1),A(NVD,1),R(1),VAL(1)
      N1=ND+1
      N2=N1+1
      NT2=NT/2
      CALL NORM (AA,VAL1,ND)
   13 NN=0
      NV=1
  102 DO 1 J=NV,N2
      IF(J-1)2,2,3
    2 DO 4 I=1,ND
    4 X(I,1)=XX(I)
      GO TO 5
    3 IF(J-N1)6,6,7
    6 DO 8 I=1,ND
      X(I,J)=XX(I)
      IF(J-I-1)8,9,8
    9 X(I,I+1)=X(I,I+1)+AA(I)/2.
    8 CONTINUE
      GO TO 5
    7 DO 10 I=1,ND
   10 X(I,N2)=A(I,1)
    5 CALL EQU(X(1,J),F(1,J))
      CALL NORM(F(1,J),VAL(J),ND)
      IF (J-1) 11,110,11
  110 VAL2=VAL(1)
      GO TO 1
   11 IF(J-N1)12,12,104
  104 IF(NN-NT2)14,14,105
  105 KK=N1
      GO TO 18
   12 KK=0
      JM=J-1
      DO 15 K=1,JM
      JK=J-K
      IF(VAL(J)-VAL(JK))17,16,16
   17 KK=K
   15 CONTINUE
   16 IF(KK)191,191,18
   18 DO 19 I=1,ND
      TEMP=X(I,J)
      TEMP1=F(I,J)
      DO 20 KI=1,KK
      JKI=J-KI
      X(I,JKI+1)=X(I,JKI)
   20 F(I,JKI+1)=F(I,JKI)
      X(I,JKI)=TEMP
   19 F(I,JKI)=TEMP1
      TEMP=VAL(J)
      DO 190KI=1,KK
      JKI=J-KI
  190 VAL(JKI+1)=VAL(JKI)
```

```
      VAL(JKI)=TEMP
191   IF(J-ND)1,1,130
130   DO 21 I=1,N1
      A(1,I)=1.
      DO 21 K=1,ND
 21   A(K+1,I)=F(K,I)
      R(1)=1.
      DO 22 K=1,ND
 22   R(K+1)=0.
      CALL CLEO(A,R,DUMMY,EPS1,N1,1,NVD,IND)
      IF(IND)33,230,33
230   DO231I=1,ND
      A(I,1)=0.
      DO231K=1,N1
231   A(I,1)=A(I,1)+X(I,K)*R(K)
      DO 499 I=1,4
499   VVV(I)=A(I,1)
      GO TO 23
 14   IF(VAL(N2)-EPS*VAL2) 30,30,310
 30   DO 32 I=1,ND
32    R(I)=X(I,1)-A(I,1)
      CALL NORM (R,VALX,ND)
      IF(VALX-EPS*VAL1)33,33,310
310   IF(VAL(N2)-VAL(N1)) 12,311,311
311   IND=-1
      IF(NN-NT2)312,312,101
312   DO 411 I=1,ND
      XX(I)=X(I,1)
411   AA(I)=X(I,2)-X(I,1)
      NV=2
      NN=NN+1
      GO TO 102
 33   DO 34 I=1,ND
 34   XX(I)=X(I,N2)
      GO TO 101
  1   CONTINUE
 23   NV=N2
      NN=NN+1
      IF(NN-NT)102,100,100
101   NT=NN
100   RETURN
       END
```

B. PROGRAM FOR CALCULATION OF DIFFUSE SURFACE REFLECTANCE

IBM 360 FORTRAN IV program for the calculation of diffuse surface reflectance by the Monte Carlo technique (Egan and Hilgeman, 1978a).

```
C     BEAM REFLECTANCE BY MONTE CARLO TECHNIQUE
      COMPLEX * 8 EM
      DIMENSION HIS(10),DELTH(10)
      REAL * 4 EENNAA(19)
      DATA EENNAA/2.,1.87,1.72,1.58,1.51,1.40,1.29,1.2,1.12,
     11.05,.99,.93,.88,.84,.8,.69,.62,.56,.51/
      DATA PO2/1.570796/,FPI/12.56637/
      COMMON EM,THETA,ANG,EYE,CUE,CON,TST,A,EPS,CTHETA
      COMMON/NPBLK/NS,NOUT,KNT,KNTOUT
      COMMON/SUMBLK/PHI,BS,FS,TS
      COMMON/TSTSUM/BS1,FS1,TS1,BS2,FS2,TS2
      COMMON/HAT/ENE,ENA,GNU
      LLL=1
      EPS=0.
      NOUT=1
      READ(5,99)NS,HEX
   99 FORMAT(I5,Z10)
      IF(HEX.NE.0.)CALL RDMIN(HEX)
  500 READ(5,94,END=150)A,EN,FK,GNU,ENA,ENE
   94 FORMAT(6E10.0)
      STEP=PO2/10.
      DO 50 I=1,10
      FI=I
      DELTH(I)=STEP*FI
   50 HIS(I)=0.
      SI=.28
      SI=.2
      SI=.6
      SI=.5
      SI=.55
      SI=.25
      SI=.1
      SI=.65
      SI=.6
      SI=.25
      SI=.5
      SI=.45
      SI=.3
      SI=.27
      SI=.35
      SI=.4
      SI=.37
      SI=.75
      SI=.7
      ENA=0.
      ENA=1000.
      ENE=     ABS(FK)*.5
      ENE=10.*ABS(FK)
      ENE=0.
      ENA=.1
      ENA=ENA/10.
      ENE=0.
      ENE=ENE*10000.
      ENA=ENA*.1
      ENE=ENE**2
      ENE=ENE*1000.
      ENA =1.
      ENA=EENNAA(LLL)
      ENA=ENA/2.
      ENA=ENA**.3333333
      ENE=     ABS(FK)
      ENE=GNU/10000. *ENE
```

```
      A=.0025
      A=.00075
      A=.0005
      A=.00025
      A=.005
      A=.0035
      A=.00015
      LLL=LLL+1
      WRITE(6,98)NS,HEX,A,EN,FK,GNU,ENA,ENE
   98 FORMAT('1TOTAL NUMBER OF SAMPLES WILL BE',I6,' RANDOM NUMBER INITI
     1ALIZER IS',Z9//' RADIUS, N, K, NU, NA, NE',1P6E16.5//)
      WRITE(6,56)SI
   56 FORMAT('0THE PROBABILITY OF SCATTERING IN IS',1PE17.5)
      AMBDA=1./GNU
      CON = -  (FPI*FK)/AMBDA
      TST = 1.E-3
      R1 = NS
      R1 = 1./R1
      IF(R1.LT.TST)TST=R1
      WRITE(6,97)ENE,ENA,CON,TST
   97 FORMAT(' NE, NA, TRANS. CONSTANT, I TEST ',1P4E18.5///)
      R1 = -FK
      EM = CMPLX(EN,R1)
      CALL CRIT
      BS1=0.
      FS1=0.
      TS1=0.
      BS2=0.
      FS2=0.
      TS2=0.
      TS=0.
      BS=0.
      FS=0.
      KNT=0
      KNTOUT=985
      R1=A**2
      J=-1
  110 CUE=0.
      EYE=1.
      K=0
   20 X = A*RDM(DUM)
      R2 = X**2
      R3 = R1-R2
      YMAX = SQRT(R3)
      Y=A*RDM(DUM)
      IF(Y.GT.YMAX)GO TO 20
      R4 = Y**2
      R5 = SQRT(R2+R4)
      ARG=R5/A
      IF(ARG.GT.1.0)ARG=1.0
      THETA=ASIN(ARG)
      ANG=THETA
      N=10
      DO 52 I=1,9
      IF(THETA.GT.DELTH(I))GO TO 52
      N=I
      GO TO 53
   52 CONTINUE
   53 HIS(N)=HIS(N)+1.
      CALL RTAS(K,-1)
      GO TO (131,132,130,133),K
  130 CALL NP(J)
```

```
      IF(J.EQ.2)GO TO 100
      GO TO 110
  131 PHI = 2.*THETA
      CALL SUM1
      GO TO 130
  132 CALL TIN(ITST)
  140 IF(ITST.EQ.1)GO TO 139
C     WRITE(6,96)EYE,TST
   96 FORMAT('0EYE LT TST, EYE,TST ',1P2E17.5)
      GO TO 130
  139 CALL RTAS(K,1)
      GO TO (141,142,130,143),K
  141 CALL RIN(ITST)
      GO TO 140
  142 CALL TOUT
      CALL SUM1
      GO TO 130
  133 ANG = THETA
  143 FLIP = RDM(DUM)
      IF(FLIP.GT.SI)GO TO 145
      CALL SINN(ITST)
      GO TO 140
  145 CALL SUM2
      GO TO 130
  100 CONTINUE
      WRITE(6,54)(DELTH(I),I=1,10),(HIS(I),I=1,10)
   54 FORMAT('0      HISTOGRAM OF THETA'/
     2' INTERVALS',10F10.5/' SUMS      ',10F10.2)
      GO TO 500
  150 STOP
      END
      SUBROUTINE NP(J)
      COMMON/NPBLK/NS,NOUT,KNT,KNTOUT
      COMMON/TSTSUM/BS1,FS1,TS1,BS2,FS2,TS2
      COMMON/HAT/ENE,ENA,GNU
      COMMON/SUMBLK/PHI,BS,FS,TS
      IF(J.EQ.-1)WRITE(6,99)
   99 FORMAT('0NUMBER OF SAMPLES, BACK SCATTER, FRONT SCATTER, TRANSVERS
     1E SCATTER, RDM NO. INIT.'//)
      J=1
      KNT = KNT + 1
      IF(KNT .LT. NS)GO TO 110
      J = 2
  100 CALL RDMOUT(HEX)
      F=KNT
      BO=BS/F
      FO=FS/F
      TO=TS/F
      BO1=BS1/F
      BO2=BS2/F
      FO1=FS1/F
      FO2=FS2/F
      TO1=TS1/F
      TO2=TS2/F
      WRITE(6,98)KNT,BO,FO,TO,HEX
   98 FORMAT(12X,I5,1P3E17.5,Z15)
      WRITE(6,96)BO1,FO1,TO1,BO2,FO2,TO2
   96 FORMAT(1P6E18.5)
      IF(KNT.EQ.NS)WRITE(7,97)BO,FO,TO,ENA,ENE
   97 FORMAT(3F10.6,2E10.4)
      GO TO 30
```

```
  110 IF(KNT.LT.KNTOUT)GO TO 30
      IF(KNT.EQ.(KNTOUT+10))KNTOUT=KNTOUT+1000
      GO TO 100
   30 RETURN
      END
      SUBROUTINE RTAS(K,INO)
      COMPLEX * 8 EM,C1,C2,C3,C4,CRS,CRP,EM1
      COMMON EM,THETA,ANG,EYE,CUE,CON,TST,A,EPS,CTHETA
C     INO = -1 OUT TO IN,  INO = +1 IN TO OUT
      CALL HATS(XH,YH)
      EM1=EM**INO
      RAN = RDM(DUM)
      CT = COS(THETA)
      TERM = EXP(-YH/CT)
      IF(TERM.GE.RAN)GO TO 100
      K=4
   30 CONTINUE
      CALL OUT(K)
      RETURN
  100 TERM = TERM - (1.-EXP(-XH/CT))
      IF(TERM.GE.RAN)GO TO 110
      K = 3
      GO TO 30
  110 C1=SIN(THETA)*EM1
      C3 = 1.-C1**2
      C3 = CSQRT(C3)
      C4=EM1
      C2 = CT/C3
C     WRITE(6,9)C1,C2,C3,C4
    9 FORMAT('OFROM RTAS C1, C2, C3, C4'/1P8E15.5)
      C1 = C4*C2
      CRS = (C1-1.)/(C1+1.)
      CRP = (C2-C4)/(C2+C4)
C     WRITE(6,8)C1,CRS,CRP
    8 FORMAT('OC1, CRS, CRP'/1P6E15.5)
      IF(INO.EQ.1 .AND. THETA.GT.CTHETA)GO TO 120
      RP = CABS(CRP)
      RS = CABS(CRS)
      RP = RP**2
      RS = RS**2
      SCRPT = (RP + RS)/2.
      TO = (1.-SCRPT)*TERM
      TERM = TERM-TO
C     WRITE(6,7)RS,RP,SCRPT,TO,TERM,RAN
    7 FORMAT('ORS, RP, SCRPT, TO, TERM, RAN'/1P6E15.5)
      IF(TERM.GE.RAN)GO TO 120
      K = 2
      GO TO 30
  120 K = 1
      GO TO 30
      END
      SUBROUTINE TIN(ITST)
      COMPLEX * 8 EM,C1
      COMMON EM,THETA,ANG,EYE,CUE,CON,TST,A,EPS,CTHETA
      DATA PI/3.141593/
      T1 = SIN(THETA)
      C1 = EM**2 - T1**2
      C1 = CSQRT(C1)
      T2 = REAL(C1)
      TR = ATAN2(T1,T2)
      EL = 2.*A*COS(TR)
      TF = EXP(CON*EL)
```

```
      EYE = EYE*TF
      IF(EYE.GE.TST)GO TO 10
      ITST = 2
      GO TO 30
   10 ITST = 1
      ANG = PI - THETA + 2.* TR
      THETA = TR
      CUE = 0.
   30 RETURN
      END
      SUBROUTINE RIN(ITST)
      COMPLEX * 8 EM
      COMMON EM,THETA,ANG,EYE,CUE,CON,TST,A,EPS,CTHETA
      DATA PI/3.141593/
      EL = 2.*A*COS(THETA)
      TF = EXP(CON*EL)
      EYE = EYE * TF
      IF(EYE.GE.TST)GO TO 10
      ITST = 2
      GO TO 30
   10 ITST = 1
      SA = SIN(ANG)
      TWT = 2.*THETA
      EPS = PI - CUE
      CB = SA * SIN(TWT) * COS(CUE)-COS(ANG)*COS(TWT)
      SB = SQRT(1.-CB**2)
      SEPS=SIN(EPS)
      IF(SEPS.LE.5.E-7)SEPS=0.
      SCUE=(SA*SEPS)/SB
      ANG = ACOS(CB)
      CUE = ASIN(SCUE)
      CUE=-CUE
      IF(ABS(CUE).LE.5.E-7)CUE=0.
   30 RETURN
      END
      SUBROUTINE TOUT
      COMPLEX * 8 EM
      COMMON EM,THETA,ANG,EYE,CUE,CON,TST,A,EPS,CTHETA
      COMMON/SUMBLK/PHI,BS,FS,TS
      ST=SIN(THETA)
      EN2=REAL(EM)**2
      FK2=AIMAG(EM)**2
      T1=.5*(EN2-FK2)
      T2=.5*ST**2
      T3=EN2*FK2
      T4=T1+T2
      T5=SQRT(T3+(T1-T2)**2)
      DEN=SQRT(T4+T5)
      THETA=ASIN(ST*DEN)
      CP = COS(THETA)*COS(ANG) - SIN(THETA)*SIN(ANG)*COS(CUE)
      PHI= ACOS(CP)
      RETURN
      END
```

```
      SUBROUTINE SINN(ITST)
      COMPLEX * 8 EM
      COMMON EM,THETA,ANG,EYE,CUE,CON,TST,A,EPS,CTHETA
      DATA PI/3.141593/,PO2/1.570796/
      EPS = PI * RDM(DUM)
   10 THETA=PO2* RDM(DUM)
      T1 = COS(THETA)
      T2 = RDM(DUM)
      IF(T1.LT.T2)GO TO 10
      EL = 2.*A*COS(THETA)
      TF = EXP(CON*EL)
      EYE = EYE*TF
      IF(EYE.GE.TST)GO TO 20
      ITST = 2
      GO TO 30
   20 ITST = 1
      SA = SIN(ANG)
      TWT = 2.*THETA
      CB = -COS(ANG)*COS(TWT) + SA*SIN(TWT)*COS(EPS)
      SB = SQRT(1.-CB**2)
      SCUE =(SIN(EPS)*SA)/SB
      ANG = ACOS(CB)
      CUE=ASIN(-SCUE)
   30 RETURN
      END
      SUBROUTINE SUM1
      COMPLEX * 8 EM
      COMMON/TSTSUM/BS1,FS1,TS1,BS2,FS2,TS2
      COMMON/NPBLK/NS,NOUT,KNT,KNTOUT
      COMMON EM,THETA,ANG,EYE,CUE,CON,TST,A,EPS,CTHETA
      COMMON/SUMBLK/PHI,BS,FS,TS
      DATA  PO2/1.570796/
      T1=(SIN(PHI))**2*EYE
      TS=TS+T1
      TS1=TS1+T1
      IF(PHI.GT.PO2)GO TO 10
      B1=(COS(PHI))**2*EYE
      BS=BS+B1
      BS1=BS1+B1
      GO TO 30
   10 F1=(COS(PHI))**2*EYE
      FS=FS+F1
      FS1=FS1+F1
   30 CONTINUE
      CALL OUT1
      RETURN
      END
```

```
SUBROUTINE SUM2
COMPLEX * 8 EM
COMMON/TSTSUM/BS1,FS1,TS1,BS2,FS2,TS2
COMMON/NPBLK/NS,NOUT,KNT,KNTOUT
COMMON EM,THETA,ANG,EYE,CUE,CON,TST,A,EPS,CTHETA
COMMON/SUMBLK/PHI,BS,FS,TS
T1 = COS(ANG)
T2 = EYE/8.
BP=T2*(1.+T1)**2
BS=BS+BP
BS2=BS2+BP
FP=T2*(1.-T1)**2
FS=FS+FP
FS2=FS2+FP
TP=EYE/4. * (3.-T1**2)
TS=TS+TP
TS2=TS2+TP
CALL OUT2
RETURN
END
SUBROUTINE HATS(XH,YH)
COMPLEX * 8 EM
COMMON EM,THETA,ANG,EYE,CUE,CON,TST,A,EPS,CTHETA
COMMON/HAT/ENE,ENA,GNU
XH = ENE
YH = ENA
RETURN
END
SUBROUTINE OUT(K)
COMPLEX * 8 EM
COMMON/NPBLK/NS,NOUT,KNT,KNTOUT
COMMON EM,THETA,ANG,EYE,CUE,CON,TST,A,EPS,CTHETA
EPS=0.
RETURN
END
```

```
      SUBROUTINE OUT1
      COMMON/NPBLK/NS,NOUT,KNT,KNTOUT
      COMMON /SUMBLK/PHI,BS,FS,TS
      RETURN
      END
      SUBROUTINE OUT2
      COMPLEX * 8 EM
      COMMON/NPBLK/NS,NOUT,KNT,KNTOUT
      COMMON EM,THETA,ANG,EYE,CUE,CON,TST,A,EPS,CTHETA
      COMMON/SUMBLK/PHI,BS,FS,TS
      RETURN
      END
      SUBROUTINE CRIT
      REAL * 4 STO(200)
      COMPLEX * 8 EM
      COMMON EM,THETA,ANG,EYE,CUE,CON,TST,A,EPS,CTHETA
      DATA TEST/0.5E-06/
      TH1=ASIN(1./REAL(EM))
      I=1
      STO(I)=TH1
   10 CALL ENTH(TH1,ENTH1)
      TH2=ASIN(1./ENTH1)
      I=I+1
      IF(ABS((TH1-TH2)/TH2).LE.TEST)GO TO 20
      STO(I)=TH2
      IF(I.GT.100)GO TO 30
      TH1=TH2
      GO TO 10
   20 CONTINUE.
      CTHETA=TH2
      STO(I)=CTHETA
      WRITE(6,9)CTHETA,I
    9 FORMAT('1SUBROUTINE CRIT CONVERGES TO',1PE17.5,' IN',I5,' STEPS'//
     1)
      WRITE(6,8)
    8 FORMAT(' THE VALUES OF THE ITERATES ARE'//)
      WRITE(6,7)(STO(K),K=1,I)
    7 FORMAT(1P5E20.5)
      RETURN
   30 WRITE(6,6)I
    6 FORMAT('1SUBROUTINE CRIT DOES NOT CONVERGE AFTER',I5,' STEPS'//)
      WRITE(6,8)
      WRITE(6,7)(STO(K),K=1,I)
      STOP
      END
      SUBROUTINE ENTH(TH1,ENTH1)
      COMPLEX * 8 EM
      COMMON EM,THETA,ANG,EYE,CUE,CON,TST,A,EPS,CTHETA
      ST=SIN(TH1)
      EN2=REAL(EM)**2
      FK2=AIMAG(EM)**2
      T1=.5*(EN2-FK2)
      T2=.5*ST**2
      T3=EN2*FK2
      T4=T1+T2
      T5=SQRT(T3+(T1-T2)**2)
      ENTH1=SQRT(T4+T5)
      RETURN
      END
```

C. PROGRAM FOR CALCULATION OF MIE SCATTERING

NOVA 800 minicomputer FORTRAN IV program for calculating Mie scattering for a sphere (after Dave, 1972).

```
MIE SCATTERING,MAIN PROGRAM
        COMPILER DOUBLE PRECISION
        DIMENSION COFA(400),FINT(185),ITIME(3)
        REAL RFR,RFI,X,QEXT,QSCAT,COFA,TEMF
   40 FORMAT(T2,'RFR=',1PD13.5,T22,'RFI=',1PD13.5,T46,'SIZE PARA
     1METER=',1PD13.5/)
   42 FORMAT(T15,'NORMALIZATION FACTOR = 4.0/(Q SUB S * X**2).')
   45 FORMAT (T10,'COEFFICIENTS OF THE LEGENDRE SERIES FOR THE SERIAL
     1 NUMBER'/)
   48 FORMAT (T2,' N ',T12,'N',T24,'N+1',T37,'N+2',T50,'N+3',T63,'N+4'
     1 )
   60 FORMAT (I5,1P5D13.5)
   62 FORMAT (/T10,'Q SUB E = ',1PD15.5)
   65 FORMAT (T10,'Q SUB S = ',1PD15.5)
   68 FORMAT (T10,'Q SUB A = ',1PD15.5)
   75 FORMAT (/T5,'INTENSITY OF THE SCATTERED RADIATION CORRESPONDING')
76        FORMAT (T5,'TO THE SCATTERIN-"KG ANGLE'/)
   78 FORMAT (/T2,'THETA',T10,'THETA',T23,'THETA + 1',T36,'THETA + 2',
     1T49,'THETA + 3',T62,'THETA + 4'/)
   80       FORMAT (0PF7.2,1P5D13.5)
        ACCEPT"TYPE 1 FOR GRAPHIC PLOT; 0 OTHERWISE",IFLG2
        IF(IFLG2.EQ.1) GO TO 3
        ACCEPT"TYPE 0 FOR PRINT OUT; 1FOR DISC STORAGE OF DATA",IFLG1
        IF(IFLG1.EQ.0) GO TO 1
3       IFLG1=1
        TYPE"DISC STORAGE OF DATA"
        GO TO 2
1       TYPE"PRINT OUT OF DATA"
2       ACCEPT"RFR,RFI,XSUB1,DELX,NX=    ",RFR,RFI,XSUB1,DELX,NX
        CALL TIME(ITIME,IER)
        TYPE"START TIME=",ITIME,"<15>"
        DO 125 I=1,185
        FINT(I)=0.0
  125 CONTINUE
      X = XSUB1
  130 CONTINUE
        CALL FOPEN(1,"T.RM")
        WRITE BINARY (1) X,RFR,RFI,QEXT,QSCAT,CTBRQS,COFA,MNCF
        CALL FCLOS(1)
        CALL FSWAP("S7.SV")
        CALL FSWAP("S2.SV")
        CALL FSWAP("S3.SV")
        CALL FOPEN(1,"T.RM")
        READ BINARY (1) QEXT,QSCAT,CTBRQS,COFA,MNCF
        CALL FCLOS(1)
        QABS=QEXT-QSCAT
        TEMA=4.0D0/(QSCAT*X**2)
      DO 160 N = 1,MNCF
      COFA(N) = TEMA * COFA(N)
  160 CONTINUE
      TEMA = 3.14159265358979332/180.0
      DO 200 I = 1,181
        TEMB = DCOS((I-1)*TEMA)
      TEMC = 1.0D+00
      TEMD = TEMB
      FINT(I) = COFA(1) + COFA(2)*TEMD
      MNCFM1 = MNCF - 1
      DO 180 N = 2,MNCFM1
      NN = N + 1
      TEME = ((2*N-1)*TEMB*TEMD - (N-1)*TEMC)/N
      FINT(I) = FINT(I) +.COFA(NN) * TEME
        TEMC = TEMD
```

```
      TEMD = TEME
  180 CONTINUE
  200 CONTINUE
        IF(IFLG1.EQ.1) GO TO 230
        WRITE(10,40) RFR,RFI,X
        WRITE(10,42)
        WRITE(10,45)
        WRITE(10,48)
        DO 150 N=1,MNCF,5
        N1=N+1
        N2=N+2
        N3=N+3
        N4=N+4
        WRITE(10,60) N,COFA(N),COFA(N1),COFA(N2),COFA(N3),COFA(N4)
150     CONTINUE
        WRITE(10,62) QEXT
        WRITE(10,65) QSCAT
        WRITE(10,68) QABS
      WRITE(10,40) RFR,RFI,X
      WRITE(10,42)
      WRITE(10,75)
        WRITE(10,76)
      WRITE (10,78)
      DO 220 I = 1,181,5
      TEMF = I - 1
        I1=I+1
        I2=I+2
        I3=1+3
        I4=1+4
        WRITE (10,80)TEMF,FINT(I),FINT(I1).FINT(12).FINT(I3).FINT(I4)
.  220 CONTINUE
        GO TO 240
230     CALL FOPEN(1,"SCATR.DA")
        WRITE BINARY(1) RFR.RFI,X,MNCF,QEXT.QSCAT,QABS
        DO 250 N=1,MNCF
        WRITE BINARY(1) COFA(N)
250     CONTINUE
        DO 260 N=1,181
        WRITE BINARY(1) FINT(N)
260     CONTINUE
        CALL FCLOS(1)
        IF(IFLG2.EQ.1) GO TO 245
        GO TO 240
245     CALL FCHAN("GRAF.SV")
240     X=X+DELX
      NX = NX- 1
      IF ( NX .GT. 0 ) GO TO 130
        CALL TIME(ITIME,IER)
        DO 1100 K=1,100
        DO 1000 M=1,100
        Y=K*K
1000    CONTINUE
1100    CONTINUE
        TYPE"END TIME =",ITIME,"<15>"
      END
```

```
SECTION S7
      COMPILER DOUBLE PRECISION
      COMPLEX RF,ACAP(400),WM1,WFN(2),RRF,TC1,TC2,FNA,FNB,RRFX,FNAP,
     1FNBP,ACAPD,FNAN,FNAD,FNBN,FNBD
      REAL FNARL(200),FNAIM(200),FNBRL(200)
     1,FNBIM(200),T(5),RX
      COMMON X,RFR, RFI,QEXT,QSCAT,CTBRQS,COFA(400)
      REAL X,RFR,RFI,QEXT,QSCAT,CTBRQS,COFA
    8 FORMAT(//T10,'THE UPPER LIMIT FOR ACAP IS NOT ENOUGH. SUGGEST, GET
     1 DETAILED OUTPUT AND MODIFY SUBROUTINE'//)
      CALL FOPEN(1,"T.RM")
      READ BINARY(1) X,RFR,RFI,QEXT,QSCAT,CTBRQS,COFA,MNCF
      CALL FCLOS(1)
      RF=DCMPLX(RFR,-RFI)
     RRF = CONJG(RF)/(RF*CONJG(RF))
     RX = 1.0D+00/X
     RRFX = RRF * RX
     T(1) = (X**2) * (RFR**2 + RFI**2)
     T(1) = DSQRT(T(1))
     NMX=1.1*T(1)
     IF ( NMX .LE. 399 ) GO TO 22
     WRITE(IO,8)
       GO TO 1001
  20 NMX2 = T(1)
     IF ( NMX1 .GT. 150 ) GO TO 22
     NMX1 = 150
     NMX2 = 135
  22 ACAP(NMX1+1) = DCMPLX(0.0D+00,0.0D+00)
     DO 23 N = 1,NMX1
     NN = NMX1 - N + 1
       ACAPD=(NN+1)*RRFX+ACAP(NN+1)
       ACAP(NN)=(NN+1)*RRFX-CONJG(ACAPD)/(ACAPD*CONJG(ACAPD))
  23 CONTINUE
     T(1) = DCOS(X)
     T(2) = DSIN(X)
       WM1=DCMPLX(T(1),-T(2))
     WFN(1) = DCMPLX(T(2),T(1))
     WFN(2) = RX * WFN(1) - WM1
     TC1 = ACAP(1) * RRF + RX
     TC2 = ACAP(1) * RF + RX
       FNAN=TC1*REAL(WFN(2))-REAL(WFN(1))
       FNAD=TC1*WFN(2)-WFN(1)
       FNA=(FNAN*CONJG(FNAD))/(FNAD*CONJG(FNAD))
       FNBN=TC2*REAL(WFN(2))-REAL(WFN(1))
       FNBD=TC2*WFN(2)-WFN(1)
       FNB=(FNBN*CONJG(FNBD))/(FNBD*CONJG(FNBD))
     QEXT = 3.0D+00 * ( REAL(FNA) + REAL(FNB))
     QSCAT = 3.0D+00 * ( REAL(FNA)**2 + AIMAG(FNA)**2 + REAL(FNB)**2
    1+AIMAG(FNB)**2)
     CTBRQS = 0.0D+00
     FNARL(1) = 1.50D+00 * REAL(FNA)
     FNAIM(1) = 1.50D+00 * AIMAG(FNA)
     FNBRL(1) = 1.50D+00 * REAL(FNB)
     FNBIM(1) = 1.50D+00 * AIMAG(FNB)
     FNAP = FNA
     FNBP = FNB
     N = 2
  65 T(1) = 2*N - 1
     T(2) = N - 1
     T(3) = 2*N + 1
     WM1 = WFN(1)
     WFN(1) = WFN(2)
```

```
      WFN(2) = T(1) * RX * WFN(1) - WM1
      TC1 = ACAP(N) * RRF + N * RX
      TC2 = ACAP(N) * RF + N * RX
        FNAN=TC1*REAL(WFN(2))-REAL(WFN(1))
        FNAD=TC1*WFN(2)-WFN(1)
        FNA=(FNAN*CONJG(FNAD))/(FNAD*CONJG(FNAD))
        FNBN=TC2*REAL(WFN(2))-REAL(WFN(1))
        FNBD=TC2*WFN(2)-WFN(1)
        FNB=(FNBN*CONJG(FNBD))/(FNBD*CONJG(FNBD))
      T(5) = N
      T(4) = T(1)/(T(5)*T(2))
      T(2) = (T(2)*(T(5)+1.0D+00))/T(5)
      QEXT = QEXT + T(3) * (REAL(FNA) + REAL(FNB))
        T(5)=REAL(FNA)**2+AIMAG(FNA)**2+REAL(FNB)**2+AIMAG(FNB)**2
      QSCAT = QSCAT + T(3) * T(5)
      CTBRQS = CTBRQS + T(2)*(REAL(FNAP)*REAL(FNA)+
     1AIMAG(FNAP)*AIMAG(FNA)+REAL(FNBP)*REAL(FNB)
     1+AIMAG(FNBP)*AIMAG(FNB))+T(4)*(REAL(FNAP)*REAL(FNBP)
     1+AIMAG(FNAP)*AIMAG(FNBP))
      T(1) = N*(N+1)
      T(1) = T(3)/T(1)
      FNARL(N) = T(1) * REAL(FNA)
      FNAIM(N) = T(1) * AIMAG(FNA)
      FNBRL(N) = T(1) * REAL(FNB)
      FNBIM(N) = T(1) * AIMAG(FNB)
      IF ( T(5) .LT. 1.0D-14 ) GO TO 100
      N = N + 1
      FNAP = FNA
      FNBP = FNB
      IF ( N .LE. NMX2 ) GO TO 65
      WRITE(10,8)
        GO TO 1001
100     CALL FOPEN(1,"T.RM")
        WRITE BINARY(1) N,RX,QEXT,QSCAT,CTBRQS
     1,FNARL,FNAIM,FNBRL,FNBIM
        CALL FCLOS(1)
        CALL FSWAP("S2.SV")
        CALL FSWAP("S3.SV")
        CALL FOPEN(1,"T.RM")
        READ BINARY(1) QEXT,QSCAT,CTBRQS,COFA,MNCF
        CALL FCLOS(1)
        CALL FOPEN(1,"T.RM")
        WRITE BINARY(1) X,RFR,RFI,QEXT,QSCAT,CTBRQS,COFA,MNCF
        CALL FCLOS(1)
        CALL FBACK
1001    TYPE"REACHED 1001"
        END
```

```
SECTION S2
       COMPILER DOUBLE PRECISION
       REAL T(2),FNARL(200),FNAIM(200),FNBRL(200)
    1,FNBIM(200),FNCKRL(400)
    1,FNCKIM(400),FNDKRL(400),FNDKIM(400)
   9 FORMAT(//T2,'THE DIMENSIONS OF FNARL,FNAIM,FNBRL AND FNBIM ARE
    1 NOT ENOUGH'/'PROGRAM TERMINATED.'//)
       CALL FOPEN(1,"T.RM")
       READ BINARY(1) N,RX,QEXT,QSCAT,CTBRQS
    1,FNARL,FNAIM,FNBRL,FNBIM
       CALL FCLOS(1)
      NMX = N
      NMXP1 = NMX + 1
      FNARL(NMXP1) = 0.0D+00
      FNAIM(NMXP1) = 0.0D+00
      FNBRL(NMXP1) = 0.0D+00
      FNBIM(NMXP1) = 0.0D+00
      T(1) = 2.0D+00 * RX**2
      QEXT = QEXT * T(1)
      QSCAT = QSCAT * T(1)
      CTBRQS = 2.0D+00 * CTBRQS *T(1)
      IF ( NMXP1 .LE. 200 ) GO TO 105
      WRITE(10,9)
        GO TO 1001
  105 K = 1
  110 FNCKRL(K) = 0.0D+00
      FNCKIM(K) = 0.0D+00
      FNDKRL(K) = 0.0D+00
      FNDKIM(K) = 0.0D+00
        I=1
  130 II = K + 2*I - 2
      IJ = II + 1
      IF ( IJ .GT. NMXP1 ) GO TO 140
      FNCKRL(K) = FNCKRL(K) + FNARL(II) - FNBRL(IJ)
      FNCKIM(K) = FNCKIM(K) + FNAIM(II) - FNBIM(IJ)
      FNDKRL(K) = FNDKRL(K) + FNBRL(II) - FNARL(IJ)
      FNDKIM(K) = FNDKIM(K) + FNBIM(II) - FNAIM(IJ)
      I = I + 1
      GO TO 130
  140 IF ( K .EQ. 1 ) GO TO 150
      T(1) = 2*K - 1
        T2=K-1
        T(2)=T2**2
      FNCKRL(K) = T(2)*FNBRL(K-1) + T(1)*FNCKRL(K)
      FNCKIM(K) = T(2)*FNBIM(K-1) + T(1)*FNCKIM(K)
      FNDKRL(K) = T(2)*FNARL(K-1) + T(1)*FNDKRL(K)
      FNDKIM(K) = T(2)*FNAIM(K-1) + T(1)*FNDKIM(K)
  150 K = K + 1
      IF ( K .LE. NMXP1 ) GO TO 110
        CALL FOPEN(1,"T.RM")
        WRITE BINARY(1) NMX,FNCKRL,FNCKIM,FNDKRL,FNDKIM,NMXP1
     1,QEXT,QSCAT,CTBRQS
        CALL FCLOS(1)
        CALL FBACK
 1001   STOP
        END
```

```
SECTION 33
         COMPILER DOUBLE PRECISION
         DIMENSION COFA(400),T(5),TA(2),FNCKRL(400),FNCKIM(400),
      1FNDKRL(400),FNDKIM(400),ITIME(3)
     7 FORMAT (//T2,'THE DIMENSION OF COFA IS NOT ENOUGH. PROGRAM
      1TERMINATED.'//)
         CALL TIME(ITIME,IER)
         CALL FOPEN(1,"T.RM")
         READ BINARY(1) NMX,FNCKRL,FNCKIM,FNDKRL,FNDKIM,NMXP1
      1,QEXT,QSCAT,CTBRQS
         CALL FCLOS(1)
         DO 140 N=1,400
         COFA(N)=0.0
140      CONTINUE
       L = 1
       T(1) = 2.0D+00
       T(2) = T(1)
       COFA(1) = 0.0D+00
       DO 170 N = 1,NMX
       T(2) = ((2*N-1)*T(2))/(2*N+1)
       NN = N + 1
       COFA(1) = COFA(1) + T(2)*(FNCKRL(NN)**2 + FNCKIM(NN)**2 +
      1 FNDKRL(NN)**2 + FNDKIM(NN)**2)
   170 CONTINUE
       COFA(1) = FNCKRL(1)**2 + FNCKIM(1)**2 + FNDKRL(1)**2 +
      1 FNDKIM(1)**2 + 0.50D+00*COFA(1)
       COFA(1) = 0.50D+00 * COFA(1)
       T(3) = 1.0D+00
   180 L = L + 2
       IF ( L .LE. 400 ) GO TO 190
       WRITE(10,7)
         GO TO 1001
   190 CONTINUE
       COFA(L) = 0.0D+00
       NR = L - 1
       NRB2 = NR/2
       IF ( NRB2 .GT. NMX ) GO TO 300
         RNR=NR
       T(1) = (4*RNR*(RNR-1)*T(1))/((2*RNR-1)*(2*RNR+1))
       T(2) = T(1)
       T(3) = ((RNR-1)**2*T(3))/(RNR**2)
       DO 250 N = NRB2,NMX
       NN = N + 1
       TA(1) = T(3)*(FNCKRL(NN)**2 + FNCKIM(NN)**2)
       TA(2) = T(3)*(FNDKRL(NN)**2 + FNDKIM(NN)**2)
       T(4) = T(3)
       DO 220 I = 1,NRB2
       IJ = 2*I
       T(4) = ((RNR-IJ+2)*(RNR+IJ-1)*T(4))/((RNR-IJ+1)*(RNR+IJ))
       II = N - I + 1
       IJ = N + I + 1
       IF ( IJ .GT. NMXP1 ) GO TO 220
       TA(1) = TA(1) + 2.0D+00 * T(4) * (FNCKRL(II)*FNCKRL(IJ)  +
      1 FNCKIM(II) * FNCKIM(IJ))
       TA(2) = TA(2) + 2.0D+00 * T(4) * (FNDKRL(II) * FNDKRL(IJ)
      1 + FNDKIM(II) * FNDKIM(IJ))
   220 CONTINUE
       COFA(L) = COFA(L) + T(2)*(TA(1) + TA(2))
       T(2) = ((2*N-RNR+1)*(2*N+RNR+2)*T(2))/((2*N+RNR+3)*(2*N-RNR+2))
   250 CONTINUE
       T(5)= NR + 0.50D+00
       COFA(L) = 0.50D+00 * T(5) * COFA(L)
```

```
      GO TO 180
 300 L = 0
       T(1)=4.0/3.0
       T(3)= 50
 310 L = L + 2
     IF ( L .LE. 400 ) GO TO 320
     WRITE(10,7)
       GO TO 1001
 320 CONTINUE
     COFA(L) = 0.0D+00
     NR = L - 1
       RNR=NR
     NRM1B2 = (NR-1)/2
     NRP1B2 = L/2
     IF ( NRP1B2 .GT. NMX ) GO TO 400
     IF ( NR .EQ. 1 ) GO TO 325
     T(1) = (4*RNR*(RNR-1)*T(1))/((2*RNR-1)*(2*RNR+1))
     T(3) = (RNR*(RNR-2)*T(3))/((RNR-1)*(RNR+1))
 325 T(2) = T(1)
     DO 350 N = NRP1B2,NMX
     TA(1)=2.0*T(3)*(FNCKRL(N)*FNCKRL(N+1)+FNCKIM(N)*FNCKIM(N+1))
     TA(2)=2.0*T(3)*(FNDKRL(N)*FNDKRL(N+1)+FNDKIM(N)*FNDKIM(N+1))
     T(4) = T(3)
     IF ( NRM1B2 .EQ. 0 ) GO TO 340
     DO 330 I = 1,NRM1B2
     II = 2 * I
     T(4) = ((II+RNR)*(II-RNR-1)*T(4))/((II-RNR)*(II+RNR+1))
     II = N - I
     IJ = N + I + 1
     IF ( IJ .GT. NMXP1 ) GO TO 330
     TA(1) = TA(1) + 2.0D+00*T(4)*(FNCKRL(II)*FNCKRL(IJ) +
    1 FNCKIM(II)*FNCKIM(IJ))
     TA(2)=TA(2) + 2.0D+00*T(4)*(FNDKRL(II)*FNDKRL(IJ) + FNDKIM(II)*
    1FNDKIM(IJ))
 330 CONTINUE
 340 CONTINUE
     COFA(L) = COFA(L) + T(2)*(TA(1) + TA(2))
     NN = 2*N
       RNN=NN
     T(2) = ((RNN-RNR)*(RNN+RNR+1)*T(2))/((RNN+RNR+2)*(RNN-RNR+1))
 350 CONTINUE
       T(5)=NR+0.5
     COFA(L) = 0.50D+00 * T(5) * COFA(L)
     GO TO 310
 400 MNCF = 2*NMX + 1
       CALL FOPEN(1,"T.RM")
       WRITE BINARY(1) QEXT,QSCAT,CTBRQS,COFA,MNCF
       CALL FCLOS(1)
       CALL TIME(ITIME,IER)
       CALL FBACK
1001   STOP
       END
```

REFERENCES

Abeles, B., and Gittleman, J. I. (1976). *Appl. Opt.* **15**, 2328.

Abraham, M., and Becker, R. (1950). "The Classical Theory of Electricity and Magnetism." Blackie, Glasgow and London.

Adams, J. B., and Filice, A. L. (1967). *J. Geophys. Res.* **72**, 5705.

Adams, J. B., and Jones, R. L. (1970). *Science* **167**, 737.

Adams, J. B., and McCord, T. B. (1969). *J. Geophys. Res.* **74**, 4851.

Allen, C. W. (1963). "Astrophysical Quantities." University of London, The Athlone Press, London.

Allen, R. M. (1962). "Practical Refractometry." R. P. Cargille Laboratories, Inc., New York.

American Cyanamid Company, Polymer and Chemicals Department, Bound Brook, New Jersey. (1977). Plastics Additives Bulletin No. 7-2701-3M-2/77.

Aronson, J. R., and Emslie, A. G. (1973). *Appl. Opt.* **12**, 2573.

Aronson, J. R., and Emslie, A. G. (1975). *In* "Infrared and Raman Spectroscopy of Lunar and Terrestrial Materials" (C. Karr, Jr., ed.). Academic, New York.

Barabashev, N. P. (1922). *Astron. Nachr.* **217**, 445.

Beckmann, P., and Spizzechino, A. (1963). "The Scattering of Electromagnetic Waves from Rough Surfaces." Pergamon, New York.

Bell, P. M., and Mao, K. C. (1972). Initial findings of a study of chemical composition and crystal field spectra of selected grains from Apollo 14 and 15 rocks, glasses and fine fractions. *In Lunar Sci. Conf., 3rd, Abstr.*, p. 55. Manned Spacecraft Center, NASA, Houston.

Bennett, A. L. (1938). *Astrophys. J.* **88**, 1.

Birkebak, R. C., Cremers, C. J., and Dawson, J. P. (1970). *Science* **167**, 724.

Bither, T. A., Bouchard, R. J., Cloud, W. H., Donohue, P. C., and Siemons, W. J. (1968). *Inorg. Chem.* **7** (11), 2208.

Bless, R. C., and Savage, B. D. (1970). "Ultraviolet Stellar Spectra and Ground-Based Observations" (L. Houzioux and H. E. Butler, eds.). Reidel, Dordrecht.

Bless, R. C., and Savage, B. D. (1972). *In* "The Scientific Results from the Orbiting Astronomical Observatory (OAO-2)" (A. D. Code, ed.). NASA Rep. No. SP-310 U.S. GPO, Washington, D.C.

Bless, R. C., Code, A. D., and Houck, T. E. (1968). *Astrophys. J.* **153**, 561.

Blevin, W. R., and Brown, W. J. (1961). *J. Opt. Soc. Am.* **51**, 129.

Bobrov, M. S. (1940). *Astron. Zh.* **17**, 1.

Bobrov, M. S. (1963). Optics and geometry of the matter of Saturn's rings. *Phys. Planetes Astrophys. Symp. 11th, Liege.* Institute d'Astrophysique, Cointe-Schlessin, Belgium.

Born, M., and Wolf, E. (1959). "Principles of Optics." Macmillan, New York.

Burns, R. G. (1970). "Mineralogical Applications of Crystal Field Theory." Cambridge Univ. Press, London and New York.
Byerly, W. E. (1959). "Fourier Series, and Spherical, Cylindrical and Ellipsoidal Harmonics." Dover, New York.
Chandrasekhar, S. (1960). "Radiative Transfer." Dover, New York.
Chapman, C. R. (1972). Surface Properties of Asteroids. Ph.D Thesis, Massachusetts Institute of Technology, Cambridge, Massachusetts.
Charette, M. P., and Adams, J. B. (1977). "Lunar Science VIII," p. 173. Lunar Science Institute, Houston, Texas.
Charette, M. P., Taylor, S. R., Adams, J. B., and McCord, T. B. (1977). "Lunar Science VIII," p. 175. Lunar Science Institute, Houston, Texas.
Conel, J. E. (1969). *J. Geophys. Res.* **74**, 1614.
Cook, A. F., Franklin, F. A., and Palluconi, F. D. (1973). *Icarus* **18**, 317.
Corning Glass works, Corning, New York. (1970). Glass Color Filter Catalog, CFG.
Dave, J. V. (1970a). *Appl. Opt.* **9**, 1888.
Dave, J. V. (1970b). *Appl. Opt.* **9**, 2673.
Dave, J. V. (1972). "Development of Programs for Computing Characteristics of Ultraviolet Radiation." NASA Rep. No. S80-RADTMO, Goddard Space Flight Center, Greenbelt, Maryland.
Deb, S. K. (1968). *Proc. R. Soc.* **304**, 211.
Deb, S. K. (1973). *Philos. Mag.* **27** (4), 801.
Deirmendjian, D. (1969). "Electromagnetic Scattering on Spherical Polydispersions." American Elsevier, New York.
Dollfus, A. (1955). Study of the Planets by Means of the Polarization of Their Light. Ph.D Thesis, University of Paris, Paris (NASA Tech. Transl. No. TTF-188, U.S. GPO, Washington, D.C.).
Dollfus, A. (1957). *Ann. Astrophys. Suppl.* **4**.
Dollfus, A. (1961). Polarization studies of planets. *In* "The Solar System: Planets and Satellites" (G. P. Kuiper and B. M. Middlehurst, eds.), Vol. 3, Chapter 9. Univ. of Chicago Press, Chicago.
Egan, W. G. (1969). *Icarus* **10**, 223.
Egan, W. G. (1971). *J. Geophys. Res.* **76**, 6213.
Egan, W. G., and Becker, J. (1968). *J. Opt. Soc. Am.* **58**, 1561.
Egan, W. G., and Becker, J. (1969). *Appl. Opt.* **8**, 720.
Egan, W. G., and Foreman, K. M. (1971). Mie scattering and the Martian atmosphere. *In* "Planetary Atmospheres" (C. Sagan *et al.*, eds.), p. 156. International Astronomical Union, Sauverney, Switzerland.
Egan, W. G., and Hallock, H. B. (1969). *Proc. IEEE* **57** (4), 621.
Egan, W. G., and Hilgeman, T. (1971). *Appl. Opt.* **10**, 2132.
Egan, W. G., and Hilgeman, T. (1972). *Bull. Am. Astron. Soc.* **4**, 425.
Egan, W. G., and Hilgeman, T. (1975a). *Appl. Opt.* **14**, 1137.
Egan, W. G., and Hilgeman, T. (1975b). *Astron. J.* **80**, 587.
Egan, W. G., and Hilgeman, T. (1976). *Appl. Opt.* **15**, 1845.
Egan, W. G., and Hilgeman, T. (1977a). *Appl. Opt.* **16**, 2861.
Egan, W. G., and Hilgeman, T. (1977b). *Icarus* **30**, 413.
Egan, W. G., and Hilgeman, T. (1978a). *Appl. Opt.* **17**, 245.
Egan, W. G., and Hilgeman, T. (1978b). *Nature* **273**, 369.
Egan, W. G., Hilgeman, T., and Reichman, J. (1973a). *Appl. Opt.* **12**, 1816.
Egan, W. G., Veverka, J., Noland, M., and Hilgeman, T. (1973b). *Icarus* **19**, 358.
Egan, W. G., Hilgeman, T., and Pang, K. (1975). *Icarus* **25**, 344.

Egan, W. G., Fischbein, W. L., Hilgeman, T., and Smith, L. L. (1978). Martian atmosphere modeling between 0.4 and 3.5 μm: Comparison of theory and experiment. *In Proc. Symp. on Planet. Atmos.*, p. 57. Royal Society of Canada, Ottawa.

Emslie, A. G., and Aronson, J. R. (1973) *Appl. Opt.* **12**, 2563.

Frank, N. H. (1966). "Introduction to Electricity and Optics." McGraw-Hill, New York.

Franklin, F. A. (1962). The Structure and Dynamics of Saturn's Rings. Ph.D. Thesis, Harvard University, Cambridge, Massachusetts.

Franklin, F. A. and Cook, A. F. (1965). *Astron. J.* **70**, 704.

Gehrels, T., Coffeen, T., and Owings, D. (1964). *Astron. J.* **69**, 826.

Gehrels, T., Roemer, E., Taylor, E. C., and Zellner, B. H. (1970). *Astron. J.* **75**, 186.

Geiger, H., and Scheel, K. (1928). "Handbuch der Physik," Vol. 20, p. 240. Springer-Verlag, Berlin.

Giovanelli, R. G. (1963). *Prog. Opt.* **2**, 111.

Gittleman, J. I., and Abeles, B. (1977). *Phys. Rev. B* **15**, 3273.

Greenberg, J. M. (1972). *J. Colloid Interface Sci.* **39**, 513.

Greenberg, J. M., and Hong, S. S. (1974). *Eslab Symp., June 4–7, 1974, Frascati, Italy*, invited paper.

Halajian, J. D., and Spagnolo, F. A. (1966). Photometric Measurements of Simulated Lunar Surfaces. Grumman Research Department Rep. No. RE-245 (Final Rep. NASA Contract No. NAS9-3182).

Hanel, R., Conrath, B., Hovis, W., Kunde, V., Lowman, P., Maguire, W., Pearl, J., Pirraglia, J., Prabhakara, C., and Schlachman, B. (1972). *Icarus* **17**, 423.

Hansen, J. E., and Travis, L. D. (1974). *Space Sci. Rev.* **16**, 527.

Hapke, B. W. (1963). *J. Geophys. Res.* **68**, 4571.

Hapke, B. W. (1966). *Astron. J.* **71**, 333.

Hapke, B., and van Horn, H. (1963). *J. Geophys. Res.* **68**, 4545.

Hardy, A. C., and Perrin, F. H. (1932). "The Principles of Optics." McGraw-Hill, New York.

Hargraves, R. B., Hollister, L. S., and Otalora, G. (1970). *Science* **167**, 631.

Hodgman, C. D. (1950). "Handbook of Chemistry and Physics." Chemical Rubber Publ. Co., Cleveland, Ohio.

Houck, J. R., Pollack, J. B., Sagan, C., Schaack, D., and Decker, J. A. (1973). *Icarus* **18**, 470.

Hoyle, F., and Wickramasinghe, N. C. (1962). *Mon. Not. R. Astron. Soc.* **124**, 417.

Hoyle, F., and Wickramasinghe, N. C. (1977). *Nature* **268**, 610.

Hoyle, F., Olavsen, A. H., and Wickramasinghe, N. C. (1978). *Nature* **271**, 229.

Huffman, D. R., and Stapp, J. L. (1971). *Nature* **229**, 45.

Huguenin, R. L., Adams, J. B., and McCord, T. B. (1977). "Lunar Science VIII," p. 478. Lunar Science Institute, Houston, Texas.

Hunt, G. R., and Salisbury, J. W. (1970). *Mod. Geol.* **1**, 283.

Hunt, G. R., Logan, L. M., and Salisbury, J. W. (1973). *Icarus* **18**, 459.

Hurlbut, C. S. (1959). "Mineralogy." Wiley, New York.

Irvine, W. M. (1966). *J. Geophys. Res.* **71**, 2931.

Irvine, W. M., and Pollack, J. B. (1968). *Icarus* **8**, 324.

Jackson, J. D. (1962). "Classical Electrodynamics." Wiley, New York.

Jenkins, F. A., and White, H. E. (1957). "Fundamentals of Optics." McGraw-Hill, New York.

Johnson, T. V., Mosher, J. A., and Matson, D. L. (1977). Lunar spectral units: A northern hemisphere mosaic. *In* "Lunar Science VIII," p. 514. Lunar Science Institute, Houston, Texas.

Kerr, P. F. (1959). "Optical Mineralogy." McGraw-Hill, New York.

Kienle, H. (1923). *Jahrb. Radioactiv. Electron.* **20** (1), 46.

Knacke, R. F., Gaustad, J. E., Gillett, F. C., and Stein, W. A. (1969). *Astrophys. J.* **155**, L189.

Körtum, G. (1969). "Reflectance Spectroscopy: Principles, Methods, Applications." Springer-Verlag, New York.
Kozyrev, N. A. (1974). *Astrophys. Space Sci.* **27**, 111.
Kramers, H. A. (1927). *Atti Congr. Fish., Como*, 545.
Kronig, R. de L. (1938). *Phys. Z.* **39**, 823.
Kruse, P. W., McGlauchlin, L. D., and McQuistan, R. B. (1962). "Elements of Infrared Technology." Wiley, New York.
Kubelka, P. (1948). *J. Opt. Soc. Am.* **38**, 448.
Kubelka, P., and Munk, F. (1931). *Z. Tech. Phys.* **12**, 593.
Kuiper, G. P., Cruikshank, D. P., and Fink, U. (1970). *Sky Telesc.* **39**, 14; **39**, 80(E).
Lathrop, A. L. (1966). *J. Opt. Soc. Am.* **56**, 926.
Lebofsky, L. A. (1973). Chemical Composition of Saturn's Rings and Icy Satellites. Ph.D. Thesis, Massachusetts Institute of Technology, Cambridge, Massachusetts.
Lebofsky, L. A., Johnson, T. V., and McCord, T. B. (1970). *Icarus* **13**, 226.
Lewis, J. L. (1974). *Bull. Am. Astron. Soc.* **6**, 337.
Lindberg, J. D., and Snyder, D. C. (1973). *Appl. Opt.* **12**, 573.
Lindblad, B. (1935). *Nature* **135**, 133.
Lowman, P. (1974). Private communication.
Lyot, B. (1929). *Ann. Obs. Paris, Section de Meudon* **8** (1) (NASA Tech. Transl. No. NASA TTF-187, Washington, D.C.).
Markov, A. V. (1924). *Astron. Nachr.* **221**, 65.
Mason, B. (1962). "Meteorites." Wiley, New York.
Maxwell-Garnett, J. C. (1904). *Philos. Trans. R. Soc. Lond.* **203**, 385.
Maxwell-Garnett, J. C. (1906). *Philos. Trans. R. Soc. Lond.* **205**, 237.
McCord, T. B., Adams, J. B., and Johnson, T. V. (1970). *Science* **168**, 1445.
McCord, T. B., Charette, M., Johnson, T. V., Lebofsky, L. A., Pieters, C., and Adams, J. B. (1972). *J. Geophys. Res.* **77**, 1349.
McCord, T. B., Pieters, C., and Feierberg, M. A. (1976). *Icarus* **29**, 1.
McLaughlin, D. B. (1961). "Introduction to Astronomy." Houghton Mifflin, Boston.
Melamed, N. T. (1963). *J. Appl. Phys.* **34**, 560.
Merrill, K. M., Russell, R. W., and Soifer, B. T. (1976). *Astrophys. J.* **207**, 763.
Mie, G. (1908). *Ann. Phys.* **25**, 337.
Minot, M. J. (1976). *J. Opt. Soc. Am.* **66**, 515.
Newcomb, S. (1960). "A Compendium of Spherical Astronomy." Dover, New York.
Oetking, P. (1966). *J. Geophys. Res.* **71**, 2505.
Orchard, S. E. (1969). *J. Opt. Soc. Am.* **59**, 1584.
Pang, K., and Ajello, J. M. (1976). *Bull. Am. Astron. Soc.* **8**, 482.
Pang, K., and Ajello, J. M. (1977). *Icarus* **30**, 63.
Pang, K., and Hord, C. (1973). *Icarus* **18**, 481.
Pang, K., Ajello, J. M., Hord, C. W., and Egan, W. G. (1976). *Icarus* **27**, 55.
Parker, H., Mayo, T., Birney, D., and McClosky, G. (1964). "Evaluation of the Lunar Photometric Function." Univ. of Virginia Res. Lab. for the Engineering Sciences Rep. No. AST-4015-101-64U, Charlottesville.
Pilcher, C. B., Chapman, C. R., and Lebofsky, L. A. (1970). *Science* **167**, 1372.
Pirraglia, J., Prabhakara, C., Schlachman, B., Levin, G., Straat, P., and Burke, T. (1972). *Icarus* **17**, 423.
Plass, G. N., and Kattawar, G. W. (1968). *Appl. Opt.* **7**, 415; **7**, 699.
Plass, G. N., and Kattawar, G. W. (1971). *J. Atmos. Sci.* **28**, 1187.
Platt, J. R. (1956). *Astrophys. J.* **123**, 486.
Pollack, J. B. (1975). *Space Sci. Rev.* **18**, 3.

Pollack, J. B., Summers, A., and Baldwin, B. (1973). *Icarus* **20**, 263.
Price, M. J. (1973). *Astron. J.* **78**, 113.
Ramo, S., and Whinnery, J. R. (1947). "Fields and Waves in Modern Radio." Wiley, New York.
Reichman, J. (1973). *Appl. Opt.* **12**, 1811.
Richtmyer, F. K., Kennard, E. H., and Lauritsen, T. (1955). "Introduction to Modern Physics." McGraw-Hill, New York.
Robinson, T. S. (1952). *Proc. Phys. Soc. London* **65B**, 910.
Robinson, T. S., and Price, W. C. (1953). *Proc. Phys. Soc. London* **66B**, 969.
Rowell, R. L., and Stein, R. S. (eds.) (1967). "Electromagnetic Scattering." Gordon and Breach, New York.
Schalen, C. (1939). *Uppsala Obs. Ann.* **1** (2).
Schalen, C. (1945). *Uppsala Obs. Ann.* **1** (9).
Schatz, E. A. (1967). *J. Opt. Soc. Am.* **57**, 941.
Schönberg, E. (1925). *Acta Soc. Sci. Fenn.* **50**, 1.
Schuster, A. (1905). *Astrophys. J.* **21**, 1.
Seeliger, H. (1887). *Abhandl. Bayer. Akad. Wiss. Math.-Naturw. Kl. II* **16**, 405.
Seitz, F. (1940). "The Modern Theory of Solids." McGraw-Hill, New York.
Simon, I. (1951). *J. Opt. Soc. Am.* **41**, 336.
Simpson, P. R., and Bowie, S. H. U. (1970). *Science* **167**, 619.
Slater, J. C., and Frank, N. H. (1947). "Electromagnetism." McGraw-Hill, New York.
Smythe, W. R. (1950). "Static and Dynamic Electricity." McGraw-Hill, New York.
Stebbins, J., Huffer, C. H., and Whitford, A. E. (1934). *Publ. Washburn Obs.* **15**, Part V.
Stebbins, J., Huffer, C. H., and Whitford, A. E., (1939). *Astrophys. J.* **90**, 209.
Stecher, T. P. (1965). *Astrophys. J.* **142**, 1681.
Stecher, T. P. (1969). *Astrophys. J.* **157**, L125.
Stockham, L. W., and Love, T. J. (1970). *J. Opt. Soc. Am.* **60**, 251.
Stratton, J. A. (1941). "Electromagnetic Theory." McGraw-Hill, New York.
Struve, F. G. W. (1847). "Etudies d'Astronomie Stellaire." Cited by Wickramasinghe (1970).
Tabor, H. (1951). Final Rep. Contract No. AF 61(052)-279 Cambridge Research Laboratories, Office of Aerospace Research, Cambridge, Massachusetts.
Tabor, H. (1955). *Bull. Res. Council Israel*, **5A** (2). (Reprinted *Trans. Int. Conf. Solar Energy*, Vol. II, Part I, Sec. A1-23.)
Tabor, H. (1959). *Sol. Energy* **3** (3), 8.
Tabor, H. (1962). *Sol. Energy* **6** (3), 112.
Tabor, H. (1966). *Sol. Energy* **10** (3), 111.
Tabor, H., Harris, J., Weinberger, H., and Doron, B. (1961). *U.N. Conf.*, paper E. Conf. 35/S46.
Tabor, H., Weinberger, H., and Harris, J. (1964). Surfaces of Controlled Absorptance. *Symp. Therm. Radiat. Solids, San Francisco.*
Trumpler, R. J. (1930a). *Lick Obs. Bull.* **14**, 154.
Trumpler, R. J. (1930b). *Publ. Astron. Soc. Pac.* **42**, 214.
Van de Hulst, H. C. (1946). *Rech. Astron. Obs. Utrecht* **9**, Part 1.
Van de Hulst, H. C. (1949). *Rech. Astron. Obs. Utrecht* **9**, Part 2.
Van de Hulst, H. C. (1957). "Light Scattering by Small Particles." Wiley, New York.
Van de Hulst, H. C. (1963). "A new look at multiple scattering." Tech. Rep, NASA Institute for Space Studies, Goddard Space Flight Center, New York.
Van Diggelen, J. (1959). *Rech. Astron. Obs. Utrecht* **14** (2), 1 (NASA Tech. Transl. No. TTF-209, U.S. GPO, Washington, D.C.).
Van Diggelen, J. (1965). *Planet. Space Sci.* **13**, 271.

Venable, W. H., Jr., Hsia, J. J., and Weidner, V. R. (1977). *J. Res. Natl. Bur. Stand.* **82** (1), 29.
Vincent, R. K., and Hunt, G. R. (1968). *Appl. Opt.* **7**, 53.
Volz, F. E. (1972). *J. Geophys. Res.* **77**, 1017.
Wamsteker, W. (1975). A Spectrophotometric Study of the Major Planets and Their Large Satellites. Ph.D. Thesis, State Univ. at Leiden, Leiden, Holland.
Weast, R. C., and Selby, S. M. (eds.) (1966). "Handbook of Chemistry and Physics," 47th ed. Chemical Rubber Publ. Co., Cleveland, Ohio.
Weill, D. F., McCallum, I. S., Bottinga, Y., Drake, M. J., and McKay, G. A. (1970). *Science* **167**, 635.
Wendlandt, W. W., ed. (1968). "Modern Aspects of Reflectance Spectroscopy." Plenum, New York.
Wendlandt, W. W., and Hecht, H. G. (1966). "Reflectance Spectroscopy." Wiley (Interscience), New York.
Wickramasinghe, N. C. (1967). "Interstellar Grains," Vol. 9: The International Astrophysical Series. Chapman & Hall, London.
Wickramasinghe, N. C. (1970). Extinction curves for graphite–silicate grain mixtures. *In Symp. on Ultraviolet Stellar Spectra and Related Ground Based Observations, 36th, Lunteren, Netherlands.* Springer-Verlag, New York.
Witt, A. N., and Lillie, C. F. (1973). *Astron. Astrophys.* **25**, 397.
Woiceshyn, P. M. (1974). *Icarus* **22**, 325.
Wolfe, W. L., ed. (1965). "Handbook of Military Infrared Technology." Office of Naval Res., Dept. of the Navy, Washington, D.C.
Young, R. H. (1977). *J. Opt. Soc. Am.* **67**, 520.

INDEX